Biochemical Aspects of
Evolutionary Biology

Biochemical Aspects of Evolutionary Biology

edited by
Matthew H. Nitecki

The University of Chicago Press
Chicago and London

MATTHEW H. NITECKI is curator of fossil invertebrates in the Department of Geology at the Field Museum of Natural History and is a member of the Committee on Evolutionary Biology at the University of Chicago.

The University of Chicago Press, Chicago 60637
The University of Chicago Press, Ltd., London

ISBN: 0–226–58684–7
LCN: 82–70746

Proceedings of the
Fourth Annual Spring Systematics Symposium:
Biochemical Aspects of Evolutionary Biology
Held at the Field Museum of Natural History
Chicago, Illinois, May 9, 1981
Sponsored by the Field Museum of Natural History
and the National Science Foundation

CONTENTS

CONTRIBUTORS

*Numbers in parentheses indicate the pages on which authors'
contributions begin.*

Stevan J. Arnold (173), *Department of Biology and Committee
on Evolutionary Biology, University of Chicago, Chicago,
Illinois, 60637*

Francisco J. Ayala (257), *Department of Genetics, University
of California, Davis, California, 95616*

Herbert G. Baker (131), *Department of Botany, University of
California, Berkeley, California, 94720*

Irene Baker (131), *Department of Botany, University of
California, Berkeley, California, 94720*

Henry C. Harpending (213), *Department of Anthropology, University of New Mexico, Albuquerque, New Mexico, 87131*

Lynne Houck (173), *Department of Education, Field Museum of
Natural History, Chicago, Illinois, 60605 and Department
of Biology, University of Chicago, Chicago, Illinois,
60637*

Lynn Margulis (9), *Department of Biology, Boston University,
Boston, Massachusetts, 02215*

Karl J. Niklas (29), *Division of Biological Sciences, Section
of Plant Biology, Cornell University, Ithaca, New York,
14853*

Thomas J.M. Schopf (1), *Department of the Geophysical Sciences
and Committee on Evolutionary Biology, University of
Chicago, Chicago, Illinois, 60637*

James A. Teeri (93), *Department of Biology and Committee on
Evolutionary Biology, University of Chicago, Illinois,
60637*

Richard H. Ward (213), *Department of Medical Genetics, University of British Columbia, Vancouver, British Columbia,
V6T 1W5*

PREFACE

Few symposia are comprised of systematists working on diverse taxonomic groups and those that do are either very local in scope or are dominated by botanists or zoologists. They rarely include paleontologists or anthropologists. The Field Museum Spring Systematics Symposia, established in 1978, fulfill a need for a regular dialogue among the botanical, zoological, and paleontological systematics communities.

The first two symposia were not published. The third, "Biotic Crises in Ecological and Evolutionary Time" held in May 1980, was published by Academic Press in 1981.

As the science of chemistry has advanced, much new information has accumulated about organisms which is relevant to their evolution and other aspects of their biology. This fourth Spring Systematics Symposium, "Biochemical Aspects of Evolutionary Biology" reviewed some areas of systematics, paleontology, biochemistry, ecology, evolution, and anthropology where chemical data have made or promise to make important contributions.

The National Science Foundation (grant no. DEB-8023139) and Field Museum provided financial support. Many staff members of the Museum, particularly Drs. William C. Burger, Lynne D. Houck, Robert F. Inger, and David M. Raup, were "looking before and after" this symposium. The following people were kind enough to review manuscripts: Drs. Randall S. Alberte, William C. Burger, Daniel J. Crawford, John J. Engel, Martin E. Feder, John W. Fitzpatrick, Michael J. Greenberg, John L. Hubby, Robert F. Inger, John B. Kethley, Timothy C. Plowman, David M. Raup, Peter H. Raven, Beryl B. Simpson, Montgomery W. Slatkin, Peter E. Smouse, Daniel J. Stewart, Lynn H. Throckmorton, Harold K. Voris, Michael J. Wade, David B. Wake and Larry E. Watrous. The real sappers and miners, however, were Dr. Kubet Luchterhand who tilled through many papers, and Miss Elizabeth Moore who with stamina and perseverance prepared the entire book for the camera ready reproduction. To all of these I extend my grateful thanks.

HISTORICAL APPROACHES VERSUS EQUILIBRIUM APPROACHES TO EVOLUTIONARY DATA

Thomas J. M. Schopf

Department of the Geophysical Sciences
and Committee on Evolutionary Biology
University of Chicago
Chicago, Illinois

INTRODUCTION

For a person trained in paleontology to be writing the
introduction to this book is a clear sign of the interdisci-
plinary nature of evolutionary biology (and is explicitly so as
I drafted this introduction at Cold Spring Harbor after taking
the course on cloning of eukaryotic genes). On a larger scale,
in June of this year the Systematics Association of Great
Britain convened a meeting at Cambridge on the Evolution
of the Genome. This symposium, which was reviewed in *Nature,*
Science and *Paleobiology,* attracted chiefly molecular biologists who have known for some time that they have to deal with
the historical origins of the genes on which they work, and
thus are actively searching for paleontological data on the
origin of particular taxa. The Field Museum, in choosing for
its Spring Systematics Symposium of 1981 the topic that led to
this book, continues to provide focal points for central areas
of evolutionary biology in the 1980's.

THE ISSUE AT LARGE

In sorting out the evolution of molecules, the same issues confront molecular biologists that have confronted naturalists for a century and a half. In the present book and related literature, those who focus on phylogeny naturally see the present world as a continuous extension of the past and focus attention on those aspects of the genome and of gene expression that are thought to have retained over time as much evolutionary information as possible. Alternatively, those whose chief interest is in ecological or behavioral questions see the world through glasses which preferentially transmit the wavelengths of the modern "equilibrium" setting -- and adjustments to it. It is the old question of how to sort out signals of evolutionary time from those of ecologic (or even developmental) time.

The ability to distinguish historical messages from equilibrium adaptations resides in the use of two approaches. First, one can use null hypotheses which allow one to predict the probability distribution of change and adjustment per unit time. The key aspect is to see the evolution of a particular phenomenon not as an event unto itself, but rather as one example among a population of events. Given the appropriate distribution, one can then sort out, for example, the probability that two given DNA molecules have similar nucleotide sequences because of convergent evolution (and hence are only analogous to each other) or because of common descent (and thus are homologous). Second, one can use multiple lines of evidence. Lines of common descent should be manifest in several ways.

The key to establishing an appropriate null hypothesis to distinguish patterns due to historical reasons as opposed to

equilibrium conditions is to know the rate of change. Thus
one is led to inquire what are the mechanisms for genomic
change, what are their rates of change, and hence how likely
it is that an equilibrium versus historical condition would
be maintained.

The main mechanisms now believed to account for genomic
change (and through genomic change to physiological and
behavioral change) were hardly talked about five years ago.
One cannot be sure but what there will be equally major sur-
prises in the near future. As currently understood the main
mechanisms of genomic change involve (1) movable genetic
elements, or transposing elements, which consist of DNA
segments of various lengths that excise from one place on a
chromosome and insert in another place on the same or a
different chromosome, and in the process alter the expression
of both structural genes and of developmental pathways; (2)
gene conversion that results in the "correction" of one copy
of a gene by another copy of the same gene, and thus leads to
families of genes evolving as a group over evolutionary time;
(3) unequal sister-stand exchange that by duplication, results
in the multiplication of DNA sequences and thus in very large
families of repeated genes; and (4) combinatorial methods of
joining DNA segments, and then further splicing their RNA
products, as in the immune system, where some 10^7 possible
products appear. (In addition there are mechanisms of single
base substitutions - mutation in the narrow sense).

All of these processes act to obliterate historical information. From the point of view of molecular evolution,
probably the most troubling process is gene conversion.   In
essence, gene conversion acts to negate random mutational
divergence between duplicated genes -- one copy is simply
"corrected" against the other. Virtually all genes have more

than one allele, and most genes seem to occur in more than one copy; seemingly any allele of any copy can act as template for the others. This leads to situations where recently corrected genes may share close to 100 percent nucleotide similarity, even though the mutation or duplication that initially led to their dual existence may have happened a very long time ago. The process of gene conversion can of course go on independently in different lineages. Hence careful attention will have to be given to those DNA segments that are sequenced for phylogenetic purposes. Reasons for gene conversion may be both stochastic (in the sense that mispairing is bound to occur, with correction a natural resultant) or deterministic (in the sense that local conditions may favor two or more identical copies of a particular DNA segment).

By obliterating historical information, each of these four mechanisms of genomic change can also act to promote genomic divergence. Since genomic divergence is the essence of speciation, speciation may now come to be seen as a much more rapid and more commonly occurring process than many biologists, at least, have customarily thought.

Appropriate null hypotheses to gauge evolutionary change would be useful for both prokaryotes and eukaryotes. Species of the enteric bacterium *Salmonella* (responsible for typhoid, food poisoning, etc.) have their genes arranged in a sequence which (with the exception of one inversion) is virtually identical to that which occurs in the normal intestinal inhabitor *Escherichia coli*; and yet there is virtually no recombination between them, and there is some 50 percent divergence in nucleotide sequences. As discussed elsewhere by John Roth, these genera display enormous stability in genome arrangement in the face of several mechanisms of genome rearrangement (especially gene duplication). Among eukaryotes,

stability of linkage groups across wide taxonomic distance is
a common observation, even leading to views of extreme chromosomal stability for 250 million years (*i.e.*, chromosome
banding patterns are very similar -- or identical -- in taxa
that diverged that long ago). Yet, just as in prokaryotes,
there are a large number of mechanisms of scrambling gene
order, and the question of what maintains stability in the
face of change remains.

In the absence of a null hypothesis of expected changes
per unit time, one is hard pressed to know if these examples
of genomic conservancy are the odd, random resultant in the
tail of the distribution of expected stochastic changes, or
whether one has identified a fundamental problem to be
answered in terms of special deterministic reasons for why one
order of genes along a chromosome is (in some sense) "better"
than another order. Having noticed the paradox of stability
in the face of change, one is tempted to jump right in and try
to solve it. However, in the absence of expectations regarding the amount of change anticipated, much effort may be for
naught. Separating equilibrium results from historical
patterns remains a basic issue.

THE ISSUE IN SPECIFICS

The issue at large is mirrored in the chapters of this
book. Three papers focus on a phylogenetic approach and
emphasize continuity of taxa through time. Three rely chiefly
on interpretations of a modern "equilibrium" setting. And
one explicitly blends historical and ecologic interpretations.

Ayala presents a broad summary of the several molecular
methods of obtaining phylogenetic information. In extreme
form, essentially all ecologic data are subsumed as time acts

to even out deviations from a stochastically ticking evolutionary clock -- or rather, clocks as one becomes increasingly
aware of different rates of change in different genes. Niklas'
paper is entirely within this phylogenetic framework of
following trends -- except that instead of considering specific
molecules through time, the goal is to mark steps in biochemical pathways through time. And still within the phylogenetic
framework, but yet more broadly developed, is Margulis' paper,
which reminds us of the biochemical bases for organizing all
organisms into Whittaker's five kingdoms.

At the opposite extreme from relying on historical, phylogenetic data is Harpending and Ward's paper, which concludes
that local demographic phenomena "account in an entirely
satisfactory way for local gene distributions." Carried to
the general case, every ecologic setting would account in an
entirely satisfactory way for every change in DNA, and its
manifestation in allele frequencies. Accordingly, to the
extent that a molecular clock exists, it would simply record
the stochastic changing of local environments through time,
as DNA chases after a changing ecology.

Harpending and Ward stress the connection between
population structure and social evolution among various
tribes of *Homo sapiens*. They rely on the prediction that
natural selection results in an equilibrium gene frequency,
and this same prediction is examined by Arnold and Houck with
respect to the pattern of occurrence of male pheromones. They
contrast the anticipated unique equilibrium condition of
natural selection with the condition of sexual selection in
which are anticipated "a tremendous variety of evolutionary
outcomes", most of which are "largely historical and nonadaptive". According to Arnold and Houck, the jury is still
out on whether pheromones do or do not result from sexual
selection.

An equilibrium, natural-selection driven response also
appears to be indicated by the polyphyletic convergence of CO_2
uptake in plants by what are called the C4 and CAM pathways,
as described by Teeri. This does not necessarily mean that
this response represents for the plants, the best of all
possible worlds. The alternative possibility is that
historical and chemical constraints are such that there are
a limited number of ways for plants to obtain atmospheric CO_2.
But within the realm of this world, C4 and CAM plants respond
as though each were an "adaptive peak".

Finally, both equilibrium and historical answers to
evolutionary problems explicitly come together in Baker and
Baker's discussion of nectar and pollination. The benefits
of pollination to flowers are related to the benefits to
visitors in receiving sugars and amino acids. The uniformity
of sugar ratios and of amino acids across wide taxonomic groups
indicate to Baker and Baker a phylogenetic constraint imposed
on the patterns of "environmentally produced variation in
amount and concentration of nectar."

Thus, in this collection of research papers and reviews
we have both state of the art summaries and explicit research
in several very active areas in biochemical approaches to
evolutionary biology. But more encompassing, we have
represented each of the major philosophical approaches to
solving evolutionary problems. The book does justice to
evolutionary biology by opting for pluralism rather than a
narrow monolithic approach. Each of the individual chapters
is part of a larger whole so that the strength of the book is
in the richness of the ways that different aspects of biochemical approaches to evolutionary biology are illuminated.

CHEMISTRY AND EVOLUTION: KINGDOMS AND PHYLA

Lynn Margulis

Department of Biology
Boston University
Boston, Massachusetts

Advances in chemistry, in particular biochemistry, have revolutionized our understanding of evolution in many ways: (1) by revealing the basis of gene replication and product synthesis (RNA, proteins); (2) by providing precise criteria for the construction of phylogenies; based on such measurable traits as chromatophore pigments, protein sequences, and alkaloid products; (3) by indicating the role of secondary metabolites in allelochemical interactions; and (4) by supplying an armamentarium of methods for determining homologies. In addition, the field of organic geochemistry has been helpful in reconstructing the nature of fossil communities, especially microbial ones.

Sophisticated chemical analysis coupled with ultra-structural, developmental, and genetic considerations have overthrown the dichotomous plant/animal classification. Several multikingdom systematic schemes have been suggested as replacement. The most internally consistent one is a modification of that proposed by the late R.H. Whittaker in 1959; in it one prokaryotic and four eukaryotic kingdoms are recognized: Monera, Protoctista, Fungi, Plantae, and Animalia. Monera include all prokaryotes, both unicellular and multicellular (e.g., eubacteria, fruiting bacteria, cyanobacteria, green oxygen-producing chloroxybacteria, actinobacteria, and others). Protoctista comprise all eukaryotic microorganisms (and their immediate descendants) that are not members of the plant, animal or fungal kingdoms. Thus this kingdom includes protists (eukaryotic unicellular forms) and their multicellular descendants that do not form embryos. About thirty phyla can be recognized: diatoms, ciliates, actinopods, chytrids, and so forth. Members of the

*Kingdom Fungi include haploid or dikaryotic organisms that
conjugate and form spores (zygo-, asco- or basidiospores);
they lack diploid stages and (9+2) organelles of motility at
all stages of development. Members of the Kingdom Animalia
are diploid organisms that develop from blastulas. Members
of the Kingdom Plantae are photoautotrophic organisms with
alternating haploid and diploid generations that develop from
nonblastula embryos.*

*Most metabolites and biochemicals are distributed through-
out members of the five kingdoms with no apparent evolutionary
correlations. However, some classes of compounds are restricted
to certain higher taxa. These include diaminopimelic acid;
certain hydrocarbons, lignans, and steroids; and the following
proteins: tubulins, globins, EF hand calcium-binding protein,
and some glycoproteins. Other proteins such as ferredoxins,
cytochromes, and histones are widely distributed yet highly
conserved. Amino acid sequence homologies can be recognized
in members of widely different genera. Thus these metabolites
and biochemicals are useful in the construction of phylogenies.*

EVOLUTION AND HIGHER TAXA

Advances in biochemistry have revolutionized our understanding of evolution in several ways. We have a deeper
understanding of neo-Darwinism which can be briefly summarized
as follows: organisms leave more offspring than the environment can support (*i.e.*, all organisms have a high biotic
potential); organisms reproduce their kind (*i.e.*, heritability
is high); natural selection acts at all stages in the life
cycle of all organisms (*i.e.*, biotic potential is not reached);
new variants with random selective values arise in populations
(some variation is due to inheritance of mutations). With
time, variants accumulate in the populations such that
descendants are recognizably distinct from their antecedent
populations.

The concept of natural selection was built on a firm base
established by Darwin. That selection acts to permit or
prevent gene transmission has been understood since the

emergence of the "new synthesis", neo-Darwinism, in the first
part of this century. However, the ultimate sources of
important variation, new characters that appear in populations
with no obvious antecedents, remained obscure. It is this
aspect of evolution -- the physicochemical nature of the
replicating and mutating material -- that has been clarified
greatly by the scientific revolution that led to the founding
of molecular biology.

We now understand the basis of replication; it involves
the complementarity of the DNA base sequences in the informational macromolecules that are present in all living
organisms. Replication always requires the production of
gene products directed by DNA. These gene products are
either RNA sequences complementary to those of the DNA, or
they are proteins synthesized under the direction of RNA
transcripts. Because of the conservatism of these macromolecules, even remotely related organisms share many macromolecular sequences (either in the base sequences of RNA
and DNA, or in the amino acid sequences of proteins). Thus
the major, but certainly not the only, sources of variation
are alterations in the base sequences of DNA, with direct
consequences for gene products.

A different source of inherited variation has been revealed
by molecular and cell biological studies coupled with natural
historical observations. This is the formation of hereditary
associations between two or more complete genomes together
with their protein synthetic systems that is, the formation
and subsequent coevolution or microbial symbioses. Many such
symbioses in which two or more genomes remain together for
much of their life cycle and tend to leave more offspring in
the associated than in the unassociated form, have been
documented: *e.g.*, lichens (algae and fungi); coral

(coelenterates and dinoflagellates); *Hydra viridis* (hydra and
chlorellas); legumes (plants and bacteria); *etc.* The
generation of inherited variation by microbial association
between heterologous genomes has probably been important in
the origin of at least two, and perhaps three, classes of
intracellular eukaryotic cell organelles (*i.e.,* plastids,
mitochondria, and cilia).

Between these two levels of variation (the monomer sequence
level and the hereditary association level), there are still
other major sources of variation (chromosomal variations)
upon which natural selection can act. Chromosomal sources
of variation have been recognized by cytogeneticists, but
their detailed behavior and overall importance as evolutionary
mechanisms have not been fully assessed. Chromosomal sources
of variation include the generation of polyploids, chromosomal
fusions and translocations, and karyotypic fissions. Karyotypic fissions in particular have probably been important in
the adaptive radiations in mammals such as canids, artiodactyls and catarrhines (Todd, 1970).

In summary, a list of important sources of inherited
variations in the total hereditary material of the biota of
the biosphere, would have to include: (1) gene base sequence
alterations, measured either directly in DNA or, indirectly
as changes in sequences of bases in RNA or of amino acids in
proteins (gene products); (2) formation of new genomes by the
integration of two or more heterologous and highly adapted
ones, as in the formation of hereditary associations; and (3)
alteration of the chromatin of eukaryotic organisms by changes
in karyotype or chromosomal organization. The remainder of
this paper will review the consequences of this application of
our understanding of sources of variation to questions of
higher taxa: kingdoms and phyla. References to the evidence
for the aforementioned mechanisms of generation of variation
are listed in table 1.

Table 1. Sources of variation in evolution

Mechanism	Reference
I. Macromolecular sequence alteration, single bases, duplications and deletions	
DNA	Doolittle, 1979; Mahler, 1981
RNA	Bonen & Doolittle, 1979
protein	Dayhoff, 1969 Dayhoff, 1976 Dickerson, 1980
II. Chromosomal alterations	
polyploidy	Jackson, 1976
chromatin reorganization modes (ciliates)	Raikov, 1972
karyotypic transformations (single chromosomal fusions, translocations)	Todd, 1975 Hsu & Benirschke, 1967–1975
karyotypic fissions	Giusto & Margulis, 1981; Todd, 1970
III. Association between entire genomes	
photoautotrophs and animals	Trench, 1979
intracellular microorganisms	Richmond & Smith, 1979; Frederick, 1981
parasites and symbionts	Cook et al., 1980
various	Schwemmler & Schenk, 1980

KINGDOMS AND CHEMISTRY

The traditional taxa have resulted from the division of
all life into the two kingdoms Animalia and Plantae. In this
scheme, all the heterotrophic eukaryotes except the fungi have
been considered animals (protozoa, metazoa), whereas the
autotrophic organisms and the bacteria, regardless of the
number of heterologous genomes they contain, have been considered,
with the fungi, as plants. There has been debate concerning the
status of eukaryotic microorganisms such as slime molds and
euglenids. Such forms never fell easily into the two Kingdom
System. The bacteria have been known for years to have both
photosynthetic and motile members and were traditionally
included in the plant kingdom, but they were seldom studied by
botanists *sensu stricto*. In fact, bacterial nomenclature
and classification practices are quite different from those
used for plants.

A combination of biochemical, electron microscopical, and
genetic techniques led to the now famous distinction, first
made by Edouard Chatton in 1938, that the living world is
divisible into two mutually exclusive modes of cellular
organization: the prokaryotic and the eukaryotic. Extending
this concept, R.H. Whittaker presented, as long ago as 1969,
a system of higher taxa based on five kingdoms. Although
Whittaker's scheme far better fits the molecular biology of
living organisms, the plant/animal system has persisted in
books published as recently as 1980. Official organizations
of microbiology and botany still use mutually inconsistent
practices for naming and classifying. For example, fungi and
bacteria are still placed with plants in many classifications,
and protozoa are under the jurisdiction of the International
Code of Zoological Nomenclature, even though many protozoa

are far more closely related to algae than they are to animals.
Unfortunately, Professor Whittaker died (in October 1980)
before he could see general acceptance of his logical
classification scheme. The following brief synopsis of the
five kingdom system includes only relatively new observations.
For a listing of recent literature on problems of higher taxa
see table 2.

Of the five kingdoms, one contains all known prokaryotes
and no other forms. This kingdom is Monera. It comprises:
bacteria including *Prochloron* (see below); Archaebacteria;
and Cyanobacteria. The four eukaryotic kingdoms are:
Protoctista, Fungi, Plantae, and Animalia. The Protoctista
include the eukaryotic microorganisms and their immediate
descendants and therefore vary widely in cell structure,
genetic organization, life cycle, and modes of nutrition.
Essentially, these are both unicellular and multicellular
organisms that are excluded from the three other well-defined
kingdoms.

The Kingdom Monera is the easiest and most unequivocal
of the five kingdoms to define. Its members are small and
either unicellular, colonial, or mycelial. They lack a
membrane-bounded nucleus and other organelles, such as
mitochondria and plastids; but they do contain such inclusions
as ribosomes, mesosomes (membranous extensions of the plasma
membrane probably associated with genophore segregation), and
gas vacuoles. Most motile prokaryotes are equipped with
unique, thin flagella (solid structures made of protein
flagellin), but some glide along surfaces via an unknown
mechanism. Monerans never have cilia, eukaryotic flagella, or
other structures with nine pairs of microtubules in the
familiar "9 + 2" array. Such "9 + 2" structures are called
"undulipodia" (Margulis, 1981). No mitotic processes have

Table 2. Problems of higher taxa: References to literature

Number of Kingdoms Suggested	Comments	References
2	plant/animal	Altman & Dittmer, 1972
3	protist/plant/animal	Poindexter, 1971
	monera/plant/animal	Stanier, et al., 1976; Dodson, 1971
4	monera/protoctista/plant/animal	Copeland, 1956
5	monera/protoctista/fungi/animal/plant	Whittaker & Margulis, 1978; Margulis, 1981; Margulis & Schwartz, 1981
8	cyanochlorobionta/fungi 1/fungi 2/chlorobionta/mycobionta, etc.	Edwards, 1976
13	monera/red algae/heterokonts/ euglenids/fungi/etc.	Leedale, 1974
--	discussion of fungi as plant subkingdom or kingdom	Ainsworth & Sussman, 1968
--	placement of bacteria in the Monera kingdom	Buchanan & Gibbons, 1974
--	placement of bacteria including cyanobacteria and chloroxybacteria in Monera Kingdom	Starr et al., 1981

ever been reported in these organisms, so meiosis is also
always absent.

Included among the monerans are organisms that differen-
tiate relatively complex structures, such as spores and
fruiting bodies (*e.g.*, the bacilli and myxobacteria respectively). Some of the monerans are clearly multicellular in
that the unit of growth and behavior in nature is composed
of hundreds to thousands of cells at least; examples include
many cyanobacteria, such as *Stigonema,* and fruiting bacteria,
such as *Stigmatella*. Some are obligate anaerobes that are
poisoned by tiny quantities of oxygen; others live optimally
as microaerophils, requiring less than ambient levels of
oxygen; still others are obligate aerobes and require 20% O_2
in the gas phase.

In addition to these structural features, prokaryotes
have a characteristic cell wall chemistry (*e.g.*, the
presence of peptidoglycans) but show enormous variations on
the basic patterns of metabolism. For example, photosynthesis
may use H_2, H_2S, organic compounds, or H_2O for the hydrogen
donor; the Krebs cycle may or may not be present; in
respiration, sulfate, nitrate, or oxygen may be reduced to
sulfide, nitrogen, or water, respectively; they may produce
methane, and they may derive energy from the oxidation of
ammonia, hydrocarbons, or sulfide. Sexuality, defined as the
formation of an offspring with more than one parent, is
absent in many prokaryotic genera. When sex is present, the
relative contributions of the "parents" to the genetic system
of the offspring are generally disparate, and the extent of
the inequality depends on many factors, such as the duration
of the mating and the composition of the medium upon which
the offspring organisms are permitted to grow.

Examples of monerans include all the bacteria (even the

"actinomycetes" fungus-like bacteria which are more
appropriately called actinobacteria), the blue-greens (now
generally called cyanobacteria), and *Prochloron*, the pro-
chlorophytes of Lewin (1981), which ought to be called
Chloroxybacteria, green oxygen-producing bacteria.

Prochloron, a new sort of prokaryotic oxygen producing
photosynthesizer, has recently been discovered and studied by
Prof. R.A. Lewin of the Scripps Institution of Oceanography.
Several strains (probably species), but only the single genus,
are known at present. Only a decade ago, *Prochloron* would
have seemed to be an anomalous collection of contradictory
characteristics. It is a coccoid unicellular oxygen-producing
photosynthetic microorganism but it differs from all other
oxygen producers. Unlike any blue-green known, *Prochloron*
lacks phycobiliproteins and therefore also lacks the blue-
green color of the other prokaryotic oxygen producers. It is
grass green in color. Like the chlorophytes (green algae) and
plants, *Prochloron* contains chlorophyll *a* and chlorophyll *b*
plus an array of carotenoids. The carotenoids include
β-carotene and generally resemble those found in green algae
and plants. But, the lack of a nucleus and mitotic karyo-
kinesis and the absence of any mitochondria or cellulosic
wall materials attest that *Prochloron* is not a chlorophyte.
The presence of diaminopimelic acid in a typically gram-
negative cell wall and their array of fatty acids place the
organisms squarely with the blue-greens and other prokaryotes.
Thus, although it has all the traditional photosynthetic
features of members of the plant kingdom, in the five kingdom
scheme, *Prochloron* is a unique member of the Monera.

By definition, all other species in the four remaining
kingdoms are eukaryotic; that is, they all have membrane-
bounded nuclei. Except for the presence of a nuclear membrane ,

exceptions can be found to every other generalization about
eukaryotes. Not all eukaryotes have mitochondria, plastids,
cellulosic walls, cilia, endoplasmic reticuli, or Fuelgen
staining chromosomes. Not all can pinocytose or phagocytose,
and not all divide by mitosis and demonstrate meiotic
sexuality. Thus, many eukaryotic species lack the features
normally associated with eukaryotes, although most have them.

How can the eukaryotic kingdoms be defined so that
explicit, noncontradictory categories emerge? The definitions
given here seem to be sufficient, comprehensible, and consistent
with both the observable natural history and the metabolic
features of the species placed in each group. The result, with
minor modifications, is R.H. Whittaker's (1969) five kingdom
scheme. The essential ecological notion of Whittaker is
preserved: in natural communities, plants are producers,
animals are consumers, and fungi are decomposers. Both Monera
and Protoctista include all three-producers, consumers, and
decomposers.

Members of the Kingdom Animalia are chemoheterotrophic
in that they obtain both energy and carbon by ingesting (or
in some cases absorbing) preformed organic compounds. They
are diploid organisms that develop from zygotes. The zygotes
are produced by anisogamy (sperm and egg), and develop into
blastulae. Animals are distinguished on morphological rather
than chemical grounds. Except for certain compounds, such as
snake venoms, collagen, keratin, calcium apatite-organic
complexes and melanin derivatives, animals are chemically
rather unexceptional.

Members of the Kingdom Plantae are photoautotrophic,
gaining energy from solar radiation and carbon from atmospheric
CO_2. They show alternations of haploid and diploid generations,
they develop from embryos, they are composed of multicellular

tissue capable of differentiation and growth, and they are
supported by sterile tissue. On this definition, the
Kingdom Plantae thus excludes all the algae and is restricted
to bryophytes (mosses, liverworts, and hornworts) and
tracheophytes (vascular plants: horsetails, lycopods, ferns,
gnetales, conifers, ginkgos, and flowering plants).
Metabolically, plants are very versatile. They produce an
impressive array of secondary compounds classified as
alkaloids, lignans, tanins, acetogenins, and waxes, most of
which are not synthesized by members of other kingdoms.

Fungi, like animals are nutritionally also chemohetero-
trophs. However, fungi are distinguished by obtaining
nutrition by absorption rather than by ingestion. They
excrete digestive enzymes and take in the resulting organic
molecules from the ambient medium across their chitinous
walls. Even carnivorous fungi never ingest even small
particles by phagocytosis.

Members of the Kingdom Fungi are defined on the basis of
their life cycles; they are either haploid or dikaryotic
(containing two coexisting haploid nuclei in common cytoplasm)
and lack undulipodia at all stages. They are never diploid,
nor do they ever form motile sperm or zoospores. They are
all able to propagate by spores, and they all lack embryos.
They conjugate by fusion of tubes called hyphae -- or, as in
some yeasts, by fusion of entire immotile single cells. Thus
the water molds, slime molds, and slime nets are excluded from
the Kingdom Fungi. In those fungi that show conjugation
processes, the diploid zygote nucleus never persists; it
immediately undergoes "zygotic meiosis", reestablishing the
haploid state. Usually, the products of meiotic division are
spores (ascospores or basidiospores). Most fungal metabolites
are poorly characterized, but some toxins and hallucinogens

are clearly limited to specific genera. The kingdom is comprised of zygomycotes, ascomycotes, and basidiomycotes with
a total of over 80,000 species. (The fungi imperfecti and
the lichens are classified as fungi by virtue of their
relationship to other members of the kingdom).

These descriptions permit the unambiguous assignment of
any living organism to one of the five kingdoms; furthermore,
compared with the traditional two kingdom scheme, they are
far more consistent with data on the distribution of metabolic
pathways in each of the major groups than is the two kingdom
scheme.

The Protoctista include all eukaryotic organisms not
conforming to the definitions of animal, plant or fungus.
Because most are free-living and of little commercial value,
the protoctists are morphologically and chemically the least
well known of the five kingdoms. Examples of major groups
include amoebae, ciliates, diatoms, dinoflagellates, slime
molds (both plasmodial and cellular), plasmodiophorans, hypho-
chytrids, apicomplexans (sporozoans with a characteristic
apical complex as seen in the electron microscope), cnido-
sporidians (sporozoans with polar filaments), actinopods
(acantharians, phaeodarians, polycystines, and heliozoans),
foraminiferans, chlorophytes, eustigmatophytes, chrysophytes,
and red and brown seaweeds. Whittaker defined protists as
unicellular eukaryotes and placed most of these organisms in
a more limited Kingdom (Protista) on the basis of their
unicellularity. However, it is clear that trends toward
multicellularity occur even in the most unicellular of these
groups (*e.g.*, ciliates, diatoms). Thus, I prefer the term
Protoctista, acknowledging that many are not unicellular at
all, yet neither are they animals, fungi, or plants (*e.g.*,
xanthophytes, labyrinthulids, comycotes). They are all aquatic
organisms that do not form embryos.

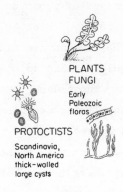

PLANTS
FUNGI

Early
Paleozoic
floras

PROTOCTISTS

Scandinavia,
North America
thick-walled
large cysts

MONERA

Warrawoona stromalites
and microfossils, Australia

Swaziland carbon and
microfossils, South Africa

ANIMALS

Ediacara
fauna

| 4.5 | 3.5 | 2.5 | 1.5 | 0.5 | 0 |

Years before present x 10^9

FIGURE 1.   *The approximate time of appearance of*
representatives of each of the five
kingdoms in the fossil record (based
on Knoll and Vidal, 1980, Pyrozinski
and Malloch, 1975 and information
summarized in Margulis, 1981)

Protoctists can also be described positively. They are
eukaryotic organisms that show staggering diversity in
mitotic and meiotic processes, chromosome structure, life
cycle, and cell organization. Conventional mitosis is lacking
in some, and meiotic sex is altogether absent in many. All
members of several major groups such as diatoms, foraminiferans
and actinopods show great complexity of cell structure. Many
protoctists are geochemically important because they produce
siliceous or calcareous tests with high preservation potential.

The five kingdom system superimposes rather well on current
concepts of the fossil history of organisms (fig. 1).

KINGDOMS AND THE FOSSIL RECORD

The relationship of the modified Whittaker Five Kingdom scheme to concepts of the origins of eukaryotic cells by symbiosis has been described in detail (Margulis, 1981). In that work, recent advances in the interpretation of the early fossil record of organisms led to the conclusion that the first several billion years of earth history were dominated by members of the Kingdom Monera. Organic geochemical analysis of well-dated carbonaceous rocks, including carbon-sulfur isotope fractionation analysis, attest to this fact (Schopf *et al.*, 1982, in press). Although most members of the Kingdoms Protoctista and Fungi fossilize poorly, it is likely that these kingdoms did not appear until long after bacterial communities in the form of stromatolites were well established. Members of the Kingdom Animalia appeared some 700 million years ago at the end of the Proterozoic Aeon (during the Vendian Era), and members of the Kingdom Plantae not until the late Silurian or Devonian Era, about 450 million years ago. The relatively late appearance of the four eukaryotic kingdoms requires a multifactored explanation. Whatever that explanation, the origin of major cell organelles and the evolution of mitotic cell division must have preceded the appearance of fungi, animals, and plants, because the cell organelles (mitotically dividing nuclei, mitochondria, and plastids -- where present) are well developed in members of all three kingdoms. The establishment of cell organelles and the stabilization of the mitotic pattern of cell division must have occurred in the adaptive radiations leading to various protoctist groups since this kingdom shows such variation in mitotic division and in the structure and composition of intracellular organelles.

Sexual reproduction such as that present in fungi, animals, plants and some protoctists requires the prior evolution of the meiotic/fertilization cycle that alternates haploidy and diploidy. The meiotic production of haploids is a modification of mitosis that required at least three steps for its evolution: (1) the diploidization of the haploid organism (for example by cannibalism or failure of cytokinesis to occur); (2) the recognition of homologous chromosomes for example, (by the formation of synaptonemal complexes insuring the segregation of homologous chromosomes, rather than sister chromatids); and (3) the retardation of kinetochore reproduction relative to chromosomal and cell division for one division cycle. The evolution of meiosis (which was probably polyphyletic) clearly preceded the emergence of the three kingdoms of large eukaryotes. The appearance of organelles and mitotic/meiotic reproductive modes occurred in the various groups of eukaryotic microorganisms, members of the Kingdom Protoctista. Even today these organisms (*e.g.*, dinoflagellates, euglenids, amoebae, ciliates) show great variations in cell structure and genetic organization. Protoctista are far more varied in these respects than animals, plants or fungi.

Many details concerning the correlation of biosynthetic ability and evolutionary history have yet to be worked out. However, the need to utilize a system of higher taxa -- kingdoms and phyla -- that most accurately represents the relationships between extant organisms remains. At present, the five kingdom system of Whittaker is apparently the most valuable, but like all taxonomies it will require continuing refinement as new data from both living and fossil forms become available.

ACKNOWLEDGMENTS

I wish to thank NASA Life Sciences and Boston University
Graduate School for research support, and Laurie Read for
aid in manuscript preparation

LITERATURE CITED

AINSWORTH, G.C., and A.S. SUSSMAN. 1968. The Fungi, An
    Advanced Treatise. New York: Academic Press.
ALTMAN, P.L., and D.S. DITTMER. 1972. Biology Data Book.
    Bethesda, Md.: Federation of American Societies for
    Experimental Biology.
BONEN, L., and W.F. DOOLITTLE. 1979. Ribosomal RNA homologies
    and the evolution of the filamentous blue-green bacteria.
    J. Mol. Evol. 10:283-292.
BUCHANAN, R.E., and N.E. GIBBONS. 1974. Bergey's Manual of
    Determinative Bacteriology. Baltimore: Williams and
    Wilkins Company.
CHATTON, E. 1938. Titres et travaux scientifique. Sète:
    E. Sottano.
COOK, C.B., P.W. PAPPAS, and E.D. RUDOLPH. 1980. Cellular
    Symbiosis and Parasitism. Columbus: Ohio State University
    Press.
COPELAND, H.F. 1956. The Classification of Lower Organisms.
    Palo Alto, California: Pacific Books.
DAYHOFF, M.O. 1969. Atlas of Protein Sequence and Structure.
    Bethesda, Md.: National Biochemical Research Foundation.
DAYHOFF, M.O. 1976. The origin and evolution of protein
    superfamilies. Fed. Proc. 35:2132-2138.
DICKERSON, R.E. 1980. Cytochrome c and the evolution of
    energy metabolism. Scientific American 242(3).
DODSON, E.O. 1971. The kingdoms of organisms. Syst. Zool.
    20:265-281.
DOOLITTLE, W.F. 1979. The cyanobacterial genome, its
    expression, and the control of that expression. Advances
    in Microbial Physiology 20:1-102. A.H. Rose and J.G.
    Morris, eds. London: Academic Press.
EDWARDS, P.E. 1976. A classification of plants into higher
    taxa based on cytological and biochemical criteria.
    Taxon 25:529-542.
FREDERICK, J.F., ed. 1981. Origin and Evolution of Eukaryotic
    Intracellular Organisms. New York: New York Academy of
    Sciences.

GIUSTO, J.P., and L. MARGULIS. 1981. Karyotypic fissioning
theory and evolution of old world monkeys and apes.
BioSystems 13:200-250.

HSU, T.C., and K. BENIRSCHKE, eds. 1967-1975. An Atlas of
Mammalian Chromosomes. Vols. 1-9. New York: Springer-
Verlag.

JACKSON, R.C. 1976. Evolution and systematic significance
of polyploidy. Ann. Rev. Ecol. Syst. 7:209-234.

KNOLL, A.H., and G. VIDAL. 1980. Late Proterozoic vase-
shaped microfossils from the Visirgsö Beds, Sweden.
Geologiska Föreningens i Stockholm Fördhandlingar (GFF)
102:207-211.

LEEDALE, G. 1974. How many are the kingdoms of organisms?
Taxon 23:261-270.

LEWIN, R.A. 1981. *Prochloron* and the Theory of Symbiogenesis.
In: Origin and Evolution of Eukaryotic Intracellular
Organelles. New York: New York Academy of Sciences.

MAHLER, H.R. 1981. Recent observations bearing on evolution
and possible origin of mitochondria. *In:* The Origins of
Life and Evolution. New York: Alan R. Liss.

MARGULIS, L. 1981. Symbiosis in Cell Evolution. San
Francisco: W.H. Freeman.

MARGULIS, L., and K. SCHWARTZ. 1981. Five Kingdoms: An
illustrated guide to the phyla of life on Earth. San
Francisco: W.H. Freeman.

POINDEXTER, J.S. 1971. Microbiology, an Introduction to
Protists. New York: Macmillan

PYROZINSKI, K. and D. MALLOCH. 1975. On the origin of land
plants: a question of mycotrophy. BioSystems 6:153-164.

RAIKOV, J.B. 1972. Nuclear phenomena during conjugation
and autogamy in ciliates. Res. Protozool. 4:147-289.

RICHMOND, M., and D.C. SMITH. 1979. The Cell as a Habitat.
London: Royal Society.

SCHOPF, J. WM., J.M. HAYES, N.W. WALTER, *et al.* 1982.
Precambrian Paleobiology Research Group, Princeton
University Press (in press).

SCHWEMMLER, W., and H.E.A. SCHENK, eds. 1980. Endocyto-
biology, Endosymbiosis and Cell Biology: A Synthesis of
Recent Research. Berlin: Walter de Gruyter.

STANIER, R.Y., E.A. ADELBURG, and J.L. INGRAHAM. 1976. The
Microbial World. Englewood Cliffs, N.J.: Prentice-Hall.

STARR, M.P., H. STOLP, H.G. TRUPER, A. BALOWS, and H.G.
SCHLEGEL, eds. 1981. The Prokaryotes: A Handbook on
Habitats, Isolation and Identification of Bacteria.
New York: Springer-Verlag.

TODD, N.B. 1970. Karyotypic fissioning and carnivore
evolution. J. Theor. Biol. 26:445-480.

TODD, N.B. 1975. Chromosomal mechanisms in the evolution of Artiodactyls. Paleobiology 1:175-188.
TRENCH, R.K. 1979. The cell biology of plant-animal symbiosis. Ann. Rev. Plant Physiol. 30:485-532.
WHITTAKER, R.H. 1959. On the broad classification of organisms. Q. Rev. Biol. 34:210-226.
WHITTAKER, R.H. 1969. New concepts of kingdoms of organisms. Science 163:150-159.
WHITTAKER, R.H., and L. MARGULIS. 1978. Protist classification and the kingdoms of organisms. BioSystems 10:3-18.

CHEMICAL DIVERSIFICATION AND EVOLUTION
OF PLANTS AS INFERRED FROM PALEOBIOCHEMICAL STUDIES[1]

Karl J. Niklas

Division of Biological Sciences
Section of Plant Biology
Cornell University
Ithaca, New York

The chemical compositions of various plant fossils, as determined by organic geochemical techniques of analyses, may be used to reconstruct some aspects of the biochemical profiles of specific taxa. Provided that these paleobiochemical data can be placed within a biosynthetic context, as inferred from modern organisms, and that a sufficiently broad range of taxa from many geological periods can be sampled, major evolutionary events at the molecular level can be determined. The relevant chemical data from such paleobiochemical studies are reviewed with regard to three major morphological events in the fossil record of plants: (1) the transition of aquatic, presumably non-vascular plants to vascular land plants (=tracheophytes), which apparently occurred during the Silurian, (2) the subsequent elaboration of vascular land plants during the early Paleozoic, and (3) the more recent advent and diversification of the flowering plants during the Cretaceous and the Tertiary. The chemical compositions of some early Silurian fossils indicate that the biosynthetic systems capable of producing lignin-like moieties and isolating these constituents in specialized cell types may have occurred during the early Silurian before the appearance of vascular tissues. With the appearance of the tracheid, at the close of the Silurian, a diverse number of compounds occur that appear to be associated with lignification, the production of cuticles, sporopollenin, and structural biopolymers. These features apparently lead to or were collateral with the adaptive radiation of tracheophytes. Paleobiochemical analyses

[1]*The author gratefully acknowledges support from the National Science Foundation, grant DEB 78-22646*

*of Devonian and early Carboniferous plants reveal the
degradative byproducts of relatively complex plant phenolics
and hydrocarbons, that permit the chemosystematic resolution
of suprageneric groups (e.g., rhyniophytes, trimerophytes,
zosterophyllophytes, and lycopods). The resolution of
taxonomic groups, by means of biochemical profiles, is seen
at its best in the characterization of more recent flowering
plant fossils. Cycloalkanes, referable to tri-, tetra-, and
pentacyclic terpanes, and such compounds as chlorophyll
derivatives, carotenoids, and flavonids (pigments which give
flowers some of their distinctive coloration) have been
isolated from Eocene and Miocene angiosperms, and permit the
assessment of chemotaxonomic relations at the species level.*

INTRODUCTION

The field of biochemical evolution is at a disadvantage
in that it has no clearly defined cognate discipline in
paleontology. Previous reconstructions of chemotaxonomic
relationships are based almost exclusively on the comparative
chemical analyses of contemporary and modern organisms, from
which patterns of evolution are deduced. Similarly, the
construction of molecular phylogenies and absolute rates
of macromolecular evolution have been obtained by comparing
the sequences in homologous molecules from relevant modern
species with inferred times of taxonomic divergence. Thus,
while molecular evolution and systematics involve the concepts
of homology and geologic times of morphologic divergence, they
lack direct integration with the fossil record. The presence
of various biosynthetic pathways and the role of referable
secondary metabolites in extinct lineages or taxa remain
conjectural. The recent development of various organic
geochemical techniques makes the chemical analysis and
characterization of fossils possible. These techniques along
with methods for efficiently and rapidly studying the
composition of many extant organisms provide for direct

comparisons among fossil and extant taxa and the reconstruction
of molecular evolutionary events. This new discipline may be
called paleobiochemistry.

The objective of this paper is to review the contributions
that paleobiochemical analyses have made to our understanding
of plant evolution. Perhaps the most detailed contributions
of paleobiochemistry have been made in the area of the more
recently evolved, terrestrial plants. This is due to the
effects of diagenesis, which tend to diminish the ability
to resolve biochemical data as a function of the age of the
fossil material. For example, Swain, F.M. *et al.* (1969) were
able to confirm the release of mono- and disaccharides from
390×10^6 year old plant tissues and infer the presence of
starch, which is diagnostic of green algae and the land
plants, while Niklas and Giannasi (1977, 1978) demonstrated
in 20×10^6 year old tissues the presence of flavonids,
diagnostic for flowering plant genera. Similarly, Prager
et al. (1980) have immunologically detected serum albumin in
40,000 year old muscle tissue from cryogenically preserved
mammoth tissues, while similar analyses of much older muscle
tissue revealed no immunologic response. While perhaps an
artifact of diagenesis, another reason for the emphasis of
relatively recent fossil material, is that the diversification
of plant secondary products appears to be most pronounced
subsequent to the appearance of terrestrial forms. Finally,
a more practical reason for emphasizing the land plant flora
is the resolution limits of the organic geochemical techniques,
which are not capable of dealing with the very small amounts
of fossil tissues often encountered when studying non-
terrestrial plants. Three major events in plant evolution
will be emphasized: (1) the origin and acquisition of land
plant characteristics, including such biochemical features

as cuticles and tracheids, (2) the diversification of the
early, vascular land plants which occurred in the Devonian,
and (3) the evolution of angiosperms. Statistical analyses
of the rates of taxonomic diversification, extinction/
origination and turnover seen for the land plants indicate
these three events coincide with major fluctuations in the
composition of the terrestrial flora (Niklas, *et al.*, 1980).
Presumably, biochemical changes coincidental with these phases
in plant evolution can be deduced and related to relevant
selective pressures.

FOSSIL PLANT BIOCHEMISTRY

Non-Vascular land plants

    Although caution must be exercised in assigning certain
organic remains to terrestrial plants (cf. Banks, 1975),
evidence is accumulating that there existed a plexus of non-
vascular plants during the early Silurian that may have been
adapted to a terrestrial environment (Pratt *et al.*, 1978;
Strother and Traverse, 1979). An increasing number of reports
indicate a widespread occurrence of spores, dyads, or tetrads,
tubular cells, and cuticle-like structures in Silurian (pre-
Pridolian) deposits. These fossils have morphological
features that are similar to those often associated with the
terrestrial plants but predate the well documented occurrence
of the earliest known vascular land plants (Edwards and
Davies, 1976; see also Edwards *et al.*, 1979; Klitzsch *et al.*,
1973). In addition to the morphological evidence, polymers
similar to macromolecules found in land plants have been
detected in early Silurian fossils (Niklas and Pratt, 1980).
These chemical data in juxtaposition with the occurrence of
banded tubular cells invite comparisons with macromolecules

(cutinic acids, polyphenols, and lignin) and water conducting
cell types found in some bryophytes and in vascular plants
(hydroids and tracheids, respectively). While the biochemical
and morphologic characters associated with some Silurian
fossils either singly or collectively do not demonstrate the
existence of vascular or even land adapted plants, their
occurrence suggests important evolutionary trends in the
acquisition of land plant features at a critical geologic
time transitional between non-vascular and vascular floras.
The biochemical changes relevant to the appearance of land
plants are critical to the reconstruction of phylogenetic
relationships among vascular and taxonomically problematic
fossil plants.

The chemical compositions of a number of early Silurian
to lower Devonian fossils are shown in table 1. For purposes
of comparison, these include data for late Silurian to early
Devonian non-vascular (*Parka*, *Pachytheca,* and *Prototaxites*) and
vascular plants (*Cooksonia* and *Psilophyton* spp). Fossils
designated as "Passage Creek Material" have been figured by
Pratt *et al.* (1978) and include parallel aligned, banded
tubes with annular to spiral ribbing, membraneous cell-like
sheets, and trilete spores, isolated from macerates of
carbonaceous siltstone from early Silurian (Llandoverian)
strata. Multicellular remains of *Eohostimella* are characterized by cylindrical, hollow axes that have been interpreted
as the remains of an erect, axial plant which was probably
adapted to the land (cf. Schopf *et al.*, 1966; Pratt *et al.*,
1978). *Parka* and *Pachytheca* are known from Pridolian to
Gedinnian strata from Europe, North America, and Siberia
(cf. Chaloner and Sheerin, 1979), and on the basis of
morphologic and biochemical information appear to be referable
to the Chlorophyta (Don and Hickling, 1917; Johnson and

Table 1.   Ultrasonic agitation extracts from representative
Silurian and early Devonian fossil plant remains

Age/Taxon	Aromatic Acids (Phenyl, $C_nH_{2n-8}O_2$)	(naphthyl, $C_nH_{2n-14}O_2$)	($C_nH_{2n-20}O_2$)
Llandoverian			
Passage Creek Material	C_{10}-$C_{14}(C_{11})^*$	C_{11}-$C_{16}(C_{16})$	--
Eohostimella	C_9-$C_{11}(C_{10})$	C_{11}-$C_{16}(C_{16})$	C_{16}-$C_{19}(C_{16})$
Pridolian			
Parka	C_8-$C_{13}(C_{10})$	$C_{17}C_{21}(C_{17})$	C_{16}-$C_{19}(C_{16})$
Pachytheca	C_7-$C_{19}(C_{12})$	C_{16}-$C_{21}(C_{17})$	C_{16}-$C_{19}(C_{16})$
Prototaxites	C_7-C_{11},$C_{14}(C_8,C_9)$	C_{11}-$C_{16}(C_{13})$	C_{15}-C_{18},$C_{20}(C_{16})$
Cooksonia	C_{13}-$C_{30}(C_{28})$	C_{14}-$C_{20}(C_{19})$	C_{15}-$C_{24}(C_{20})$
Stegenian			
Psilophyton princeps	C_{13}-$C_{30}(C_{28})$	C_{14}-$C_{20}(C_{19})$	C_{15}-$C_{24}(C_{19})$
P. cf forbesii	C_{13}-$C_{30}(C_{28})$	C_{14}-C_{18},$C_{20}(C_{20})$	C_{15}-$C_{24}(C_{19})$

[*]Determined by gas chromatography - mass spectroscopy; carbon chain-length range with maximum in parenthesis

Konishi, 1958; Niklas, 1980). *Prototaxites* is one of several
thalloid plant fossils whose structure leaves its relationship to living or other fossil forms in doubt. In the type
species, *P. loganii,* the "axis" of the plant consists of
large tube-like elements, extending generally length-wise
through a matrix of smaller-diameter, crooked tubes. While
the genus has been assigned to the Phaeophyta its affinities
remain conjectural (cf. Niklas, 1980).

Chemical analyses of the "non-vascular" plants when
juxtaposed with the compositions of known vascular plants of
comparable age such as *Cooksonia* and *Psilophyton* indicate
numerous similarities and significant differences that are
useful in determining their probable evolutionary status
toward being land/or vascular plants (table 1). p-Hydroxymono-
carboxylic acids have been isolated from *Prototaxites* (Niklas,
1980 and references therein). These constituents have been

Table 1. (cont.)

Phenols	Dicarboxylic Acids	Aromatic Dicarboxylic Acids		
$(C_nH_{2n-18}O)$	$(C_nH_{2n-2}O_4)$	$(C_nH_{2n-10}O_4)$	$(C_nH_{2n-16}O_4)$	$(C_nH_{2n-18}O_4)$
C_{10}-$C_{22}(C_{14})$	--	C_{10}-$C_{16}(C_{14})$	C_{10}-$C_{16}(C_{16})$	--
C_{10}-$C_{22}(C_{17})$	C_{12}-C_{14}	C_{10}-$C_{12}(C_{12})$	C_{10}-$C_{16}(C_{14},C_{16})$	C_{10}-$C_{14}(C_{11})$
C_{15}-C_{18}	$C_5,C_6(C_6)$	$C_8,C_{12}(C_{12})$	C_{11},C_{14}-$C_{16}(C_{14})$	C_{12}-$C_{20}(C_{18})$
--	$C_5,C_6(C_6)$	$C_8,C_{12}(C_{12})$	C_{10}-$C_{16}(C_{14})$	C_{11}-$C_{19}(C_{18})$
C_{13}-C_{15},C_{17}-$C_{20}(C_{14})$	C_4-$C_{10}(C_{15})$	C_7-$C_{12}(C_7)$	C_{12},C_{14}-$C_{16}(C_{14})$	C_{13},C_{15}-$C_{16}(C_{16})$
C_{10}-C_{25}	C_{10}-$C_{15}(C_{15})$	C_9-$C_{14}(C_{12})$	C_{10}-C_{16}	C_9-C_{14}
C_{10}-$C_{26}(C_{25})$	C_{11}-$C_{16}(C_{14})$	C_{11}-$C_{14}(C_{12})$	C_{10}-C_{16}	C_9-$C_{17}(C_{14})$
C_8-$C_{17},C_{26}(C_{26})$	C_{11}-$C_{16}(C_{14})$	C_{11}-$C_{15}(C_{12})$	C_{16}	C_9-C_{17}

isolated from modern plant cuticle and from fossil vascular plants (Matic, 1956; Niklas and Chaloner, 1976). Monohydroxy acids with carbon chain-lengths of C_{18} - C_{22}, isolated from *Prototaxites*, may be interpreted as derivatives of suberin-like biopolymers. In addition, aliphatic di- and trihydroxy acids in the form of their polyesters and in large concentrations have been reported for *Prototaxites* (Niklas, 1980). These constituents function in modern plants as components of cuticle where they function in preventing desiccation. While possibly serving a similar function, these chemicals are not components of a *bona fide Prototaxites* cuticle, since an external epidermal-like layer of cells surrounding the tubular construction of this genus has never been documented. The data available for *Parka* and *Pachytheca* are consistent with a green algal interpretation (Niklas, 1980). Gas chromatograms of both the reproductive and vegetative portions of *Parka* reveal a series of phenyl- and naphthylaromatic acids, phenols, and dicarboxylic acids (table 1) as well as sterane derivatives (stigmastane and ergostane), which are charac-teristic of many modern green algae (Niklas, 1980). These chemical data are consistent with the pseudoparenchymatous construction of *Parka* which is similar to that seen in some modern representatives of the Chaetophorales (*e.g.*, *Phycopeltis*, *Coleochaete*). In addition to this similarity, both *Parka* and *Phycopeltis* have been shown to produce sporopollenin in their spore walls (Good and Chapman, 1978; Niklas, 1980).

Perhaps the most intriguing set of biochemical constituents isolated from Silurian plant remains have been obtained by the oxidative degradation of fossils with alkaline nitro-benzine or by heating the remains with aqueous $NaOH-Na_2S$ and then methylating and oxidizing the solubilized material ($KMnO_4-NalO_4$ in aqueous NaOH, then H_2O_2 in aqueous Na_2CO_3). Both *Eohostimella* and the Passage Creek fossils contain phenolic cell wall constituents that are very similar to those released by lignin when it is oxidized (cf. Niklas and Pratt, 1980). Gas-chromatography reveals the presence of p-hydroxy-benzaldehyde, vanillin, small amount of syringaldehyde, and vanillic and syringic acids. High vacuum pyrolysis (at 450 C) of the Passage Creek fossils, native lignin (from modern spruce), and axes from Devonian *Psilophyton* reveal similar patterns indicating the common presence of a variety of constituents (fig. 1; table 2). A large proportion of the compounds released at 450 C are CO_2 and low molecular weight alkanes and alkenes. The dominant aromatics consist of phenol, o-methyl-phenyl (cresol), and dimethylbenzofurans. These data are similar to those reported for spruce lignin and for silicified conifer wood (*Araucarioxylon arizonicum*)

Table 2. Pyrolysis Composition of Lignin and Lignin-like Compounds.

Compound	No.	Spruce Lignin	Callixylon	Passage Creek Material	Polytrichum commune
benzene	1	+	+	+	+
toluene	2	+	+	+	-
ethylbenzene	3	+	+	+	-
p-xylene	4	+	-	+	-
m-xylene	5	+	+	+	-
o-xylene	6	+	+	+	-
styrene	7	+	+	+	-
ethyltolicene	8	+	+	+	-
trimethylbenzene	9	+	+	+	-
benzofuran	10	+	+	+	+
indene	11	+	+	+	-
2-methylbenzofuran	12	+	+	+	+
phenol	13	+	+	+	+
o-cresol	14	+	+	+	-
dimethylbenzofuran	15	-	-	+	+
naphthalene	16	+	+	+	+
trimethylindane	17	-	-	+	-
naphthalene	18	-	-	+	-
methylxylenols	19	-	-	+	-
1-methylnaphthalene	20	-	-	+	-
7-methyl benzofuran	a	+	+	-	-
trimethyl-1-idanone	b	+	+	-	-
5 methyl-2-furfurylfuran	c	-	+	-	-

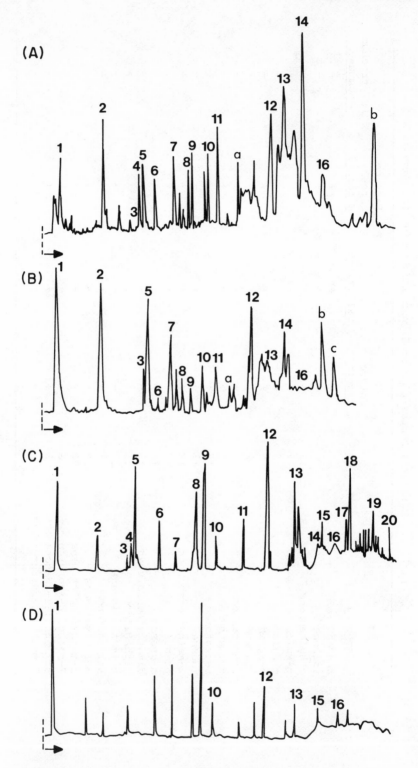

from the Triassic Chinle Formation (Sigleo, 1978).

While chemical similarities exist between some of the
components isolated from the Silurian Passage Creek fossils
and from native spruce lignin, various non-vascular plants are
known to produce complex polyphenols. Miksche and Yasuda (1978)
have reported methyl esters obtained by oxidative degradation
from some "giant" mosses of the Polytrichales. These esters
include methyl veratrate, dimethyl isohemipate, and dimethyl
metahemipate which could be derived from lignin. This origin
was excluded, however, since both 3', 4,5-trimethoxy-3,4'-
oxydibenzoate and dimethyl 5,5'-dehydro-diveratrate are absent
from the extract. These constitutents are always obtained
from lignins derived from coniferyl alcohol. The 450 C
pyrolytogram obtained from the moss *Polytrichum commune*
(gametophytic axes) shows some similarities with those
obtained from modern and fossil lignin and the Passage Creek
specimens (fig. 1). The presence of o-methoxyphenol (guaiacol),
4-methyl-2-methoxyphenol, and 4-ethyl-2-methoxyphenol in the
fossil and modern vascular tissue is consistent with their
known lignin composition. Trace amounts of some of these
compounds are detectable in the Passage Creek material and are
lacking in the moss tissues examined. Extreme care must be
taken, however, to determine if the lignin-like constituents
isolated are the altered products of lignin or polyphenolic

FIGURE 1. *Gas-chromatography pyrolysis (450 C) of native*
lignin extracted from spruce wood (A), Psilophyton *axes (B), microfossils from the Silurian*
(Llandoverian) Passage Creek locality (C), and
gametophyte axes of the moss Polytrichum commune
(D). Mass-spectroscopy identification of the
peaks, as well as a summary distribution of
these compounds in the plants examined are
given in Table 2.

constituents similar to those produced by modern mosses. The
difficulty in interpreting data such as these is compounded
since lignin isolated from modern plants may vary considerably
in its monomeric composition. Principle component analyses
provide a statistical technique whereby the influence of
diagenetic phenomena on fossil taxa can be directly compared.
An analysis, comparing the chemical compositions of *Eohostimella* and the Passage Creek fossils with Silurian and
Devonian vascular plants (*Cooksonia* and *Psilophyton* spp.) and
fossil (*Parka, Pachytheca*) and extant (*Dawsonia* and *Dendro-
ligotrichum*) thallophytes, is shown in fig. 2. The Silurian
Eohostimella and the Passage Creek fossils cluster with
Cooksonia. Two specimens referable to presumed species of
the Devonian vascular plant *Psilophyton* appear distinct from
the *Cooksonia* grouping. Similarly, fossil and extant thallophytes form distinct clusters. On the basis of this method
of comparison, *Eohostimella* and the Passage Creek fossils are
chemically most similar to the earliest known vascular plant
Cooksonia. Discrepancies between the compositions of Silurian
plant lignin-like constituents and "true" lignin may be the
result of extreme diagenesis. Yet another interpretation is
that the biosynthetic mechanism associated with lignin formation in modern plants had not occurred even by the time
Cooksonia evolved. Evidence for this may be seen in the
differences between the lignin reported for various rhyniophytes and trimerophytes (cf. Niklas and Gensel, 1976). The
integration of a lignin biosynthetic pathway with the
developmental processes associated with vascular tissue
formation may not have occurred in the early Silurian and
may have not culminated until the late Silurian-early Devonian.
The occurrence of plant lineages in which part, but not all,
of the developmental/biosynthetic machinery necessary for

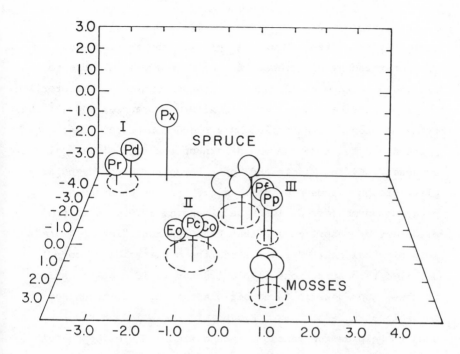

FIGURE 2. *Principle component analyses of fossil green
algae (cluster I), Prototaxites (Px), Silurian
fossils (cluster II), Devonian vascular plants
(cluster III), and modern spruce wood and moss
gametophytes (symbols for taxa are given in
Table 3). Silurian plant remains found in
strata thought to be derived from terrestrial
sediments (Eo, Pc) cluster with the known
vascular plant* Cooksonia *(Co), but are distinct
from modern and fossil vascular tissues, as well
as modern moss axes.*

vascular tissue formation had evolved may account for a number
of fossils for which a taxonomic affinity remains problematical.
Until additional data are available on the lignin-like
constituents found in some Silurian fossils the precise phylogenetic implications must remain unresolved.

Early Vascular Land Plants

Three suprageneric groups of Lower Devonian plants are
currently recognized (Banks, 1968): (1) the rhyniophytes,
characterized by plant axes that bear terminal, globose to
ellipsoid sporangia, (2) the zosterophyllophytes, characterized
by plant axes bearing lateral, reniform sporangia, and (3) the
trimerophytes, characterized by a pseudomonopodial branching
pattern and fusiform sporangia born on the tips of lateral
branches. By the Middle and Upper Devonian, other vascular
plant groups become prominent, *e.g.*, the lycopods and
progymnosperms (cf. Chaloner and Sheerin, 1979). With the
discovery of fossil genera that have apparent "intermediate"
morphological characteristics and the publication of more
detailed reconstructions of plants, the rigid classification
of these suprageneric Paleozoic plant groups is recognized
to be in a state of flux. Recently, chemical data have been
used to help define taxonomic boundaries and provide useful
criteria which are, at least superficially, independent of
morphologic and anatomic features (cf. Niklas, 1979, 1980).
The states of preservation, geologic age, and the locality
data available for a number of fossil plants are summarized
in table 3, while the referable chemical data available for
these fossils are given in table 4. The fossil plants
selected include representatives of the rhyniophytes, zosterophyllophytes, trimerophytes, and lycopods, fossils thought to
be green algae (*e.g.*, *Botryococcus, Parka, Pachytheca*), and
algal-like organisms (*e.g.*, *Prototaxites, Spongiophyton,
Nematothallus*). While more detailed information on compounds
apparently unique for each taxon is given by Niklas (1980,
and references therein), qualitative differences in chemical
compositions are summarized by means of statistical topologies
as determined by the simple matching coefficient (S_{sm}), which

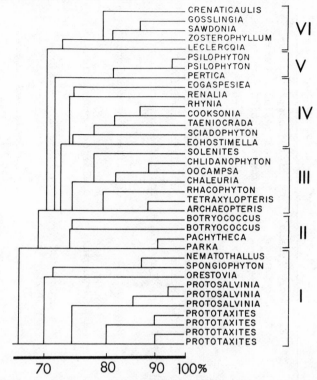

FIGURE 3.   *Differences in chemical compositions of various
vascular and non-vascular plant groups expressed
as percent of similarity (bottom axis). Statistical topology, is determined by the simple
matching coefficient (S_{sm}) method. Fossils
interpreted to be referable to distinct suprageneric chemical profiles are grouped into six
clusters designated by roman numerals.*

yields a dendrogram (fig. 3). The dendrogram is derived from

the TAXAN 5 computer program package and represents a basic

data matrix (BDM) of 36 fossils by 52 chemical character states

(table 3-5). On the basis of this technique, six groupings or

clusters of fossils are discernable. These are designated by

roman numerals I-VI (fig. 3). Cluster I represents five non-

vascular genera which, on the basis of their chemical com-

positions, appear to be algal in nature. Some of these fossils

TABLE 3. PALEOBOTANICAL DATA

Genus	Symbol	Age	Locality	State of preservation
Archaeopteris	A	Lower Frasnian	Ellesmere Island, Arctic Canada	Coalified compressions- partial cellular per- mineralization
Botryococcus	B	Upper Mississip- pian to present	Alaska, Australia England, Europe, North America	Ergastic sheath
Chaleuria	C	Lower or Middle Devonian	Chaleur Bay, New Brunswick	Compression, Abundant coalified material
Chlidanophyton duplinenis	Ch	Lower Mississip- pian	Dublin, Virginia	Coalified compression
Cooksonia sp.	Co	Middle-Upper Ludlovian	Bohemia	Coalified compressions
Crenaticaulis verruculosus	Cr	Lower Devonian	Gaspe Peninsula	Coalified compressions
Drepanophycus cf. spinaeformis	D	Upper Devonian	New Brunswick, New Jersey	Coalified compression
Eogaspesieae gracilis	Eo	Lower Devonian	Gaspe, Canada	Highly coalified compres- sion
Eohostimella	E	Lower Silurian	Maine, U.S.A.	Cylinders of coalified tis- sues (?Cuticle, cortex, vascular)

Table 3 (cont.)

Genus	Symbol	Age	Locality	State of preservation
Gosslingia breconensis	G	?Gedinnian-Siegenian Boundary	Brecon Beacons, South Wales	Coalified compressions
Leclercquia complexa	L	Middle Givetian	Brown Mt., New York, Gilboa	Coalified compressions
Lepidodendropsis ?scobiniformis	Lp L'p	Lower Mississippian	Dublin, Virginia	Highly coalified compression
Nematothallus	N	Upper Silurian to ?Upper Devonian	Freshwater Est, England	Cellular reticulum seen on surface of cuticles
Oocampsa cathea	Oo	Early Middle Devonian	New Brunswick	Coalified compressions
Orestovia	O	Lower Devonian	Kuznetzk Basin, Yunnan, China	
Pachytheca	Pd	Upper Silurian to Lower Devonian	England, Wales, Scotland, Nova Scotia	Filamentous patterns seen in carbonified material
Parka	Pr	Upper Silurian to Lower Devonian	England-Wales, Scotland, Nova Scotia	Pseudoparenchymatous cellular pattern, very thin

Table 3. *(cont.)*

Genus	Symbol	Age	Locality	State of preservation
Pertica	Pt	Late Lower Devonian	Battery Point Formation, Gaspe, Quebec	Compression. Abundant organic material
Protosalvinia	Ps	Upper Devonian	East-Central U.S.A., Middle Amazon Basin, Brazil	Thick cuticle, showing external and internal patterns of cells
Prototaxites	Px	Upper Silurian to Upper Devonian	Scotland, Wales Canada	Organic material in the form of hyphal-like tubes
Pseudosporochnus nodosus	Psp	Lower Givetian	Belgium	Coalified compressions
Psilophyton spp.	Pr Pf	Lower Devonian	Battery Point Formation, Gaspe Sandstone, Canada	Compression-impression. Organic material sparse.
Renalia	R	Lower Devonian	Arostok County, Maine, U.S.A.	Compression. Organic material fairly abundant, carbonized
Rhacophyton cerantangium	R	Upper Devonian	Valley Head, West Virginia	Partially pyritized compressions

Table 3. (cont.)

Genus	Symbol	Age	Locality	State of preservation
Rhynia spp.	Ry	Lower Devonian (?Siegenian-Esmian)	Aberdeenshire, Scotland	Cellular permineralization
Sawdonia	S	Middle Devonian	Battery Point Formation, Gaspe Sandstone, Canada	Compression-impression. Organic material sparse.
Sciadophyton ?laxum	Sc	Lower Devonian	Seal Rock Ledge, Gaspe, Canada	Coalified compression
Solenites	So	Jurassic	Yorkshire, England	Compression. Leaf cuticle
Spongiophyton	Sp	Middle Devonian	Parana, Brazil, Ghana	Thick, with pores (on one surface generally), inner patterns of cells
Taeniocrada	Ta	Lower to Middle Devonian	Germany, Belgium, Main, U.S.A.	Cellular reticulum occasionally seen, ?central vascular strand
Tetraxylopteris Schmidtii	T	Upper Givetian	Green County, New York	Coalified compressions-partial cellular permineralization
Triphyllopteris	Tr	Lower Mississippian	Pulaski, Virginia	Coalified compression
Zosterophyllum	Z	Devonian	Bulgavies Quarry, Penhallow, Scotland	Carbonaceous compression

Table 4. Range of Chemical Compounds Found in Fossils

	Normal acids	Aromatic Acids phenyl $(C_nH_{2n-8}O_2)$	Aromatic Acids naphthyl $(C_nH_{2n-14}O_2)$	Phenols $(C_nH_{2n-18}O)$
Archaeopteris	C_9-C_{33}	C_9-C_{30}	C_{10}-C_{33}	C_8-C_{28}, C_{30}-C_{31}
Botryococcus	C_7, C_9-C_{28}	C_7-C_{18}	C_{13}, C_{14}	C_{12}-C_{20}
Chaleuria	C_9-C_{38}	C_9-C_{29}	C_9-C_{29}	C_9-C_{29}
Chlidanophyton	C_{10}, C_{12}-C_{33}	C_8-C_{31}	C_{12}, C_{14}-C_{33}	C_9-C_{30}
Cooksonia	C_9-C_{29}, C_{31}	C_9-C_{30}	C_9-C_{30}	C_{10}-C_{25}
Crenaticaulis	C_8-C_{28}	C_7-C_{20}	C_9-C_{21}	C_9-C_{28}
Drepanophycus	C_8-C_{28}	C_{12}-C_{31}	C_{11}-C_{33}	C_9-C_{23}
Eogaspesiea	C_8-C_{28}	C_5, C_7-C_{21}	C_8-C_{21}	C_{10}-C_{22}
Eohostimella	C_5, C_7-C_{26}, C_{28}	C_8-C_{14}	C_8-C_{21}	C_{10}-C_{22}
Gosslingia	C_8-C_{26}, C_{28}	C_7-C_{20}	C_9-C_{20}	C_{11}-C_{28}, C_9-C_{28}
Leclercqia	C_8-C_{28}	C_7-C_{20}	C_9-C_{22}	C_9-C_{23}
Lepidodendropsis	C_8-C_{28}	C_{12}-C_{31}	C_{11}-C_{33}	C_9-C_{24}
Namatothallus	C_8-C_{36}	C_7-C_{12}	C_{11}-C_{18}	C_{12}-C_{20}
Oocampsa	C_9-C_{35}	C_9-C_{30}	C_{10}-C_{30}, C_{33}	C_8-C_{22}
Orestovia	C_6, C_8-C_{26}	C_6, C_8-C_{20}	C_{12}-C_{18}	C_9-C_{18}
Pachytheca	C_7-C_{21}	C_7-C_{13}	C_{17}-C_{21}	C_{12}-C_{20}
Parka	C_8-C_{21}	C_8-C_{13}	C_{17}-C_{21}	C_{13}-C_{18}
Pertica	C_9-C_{35}	C_8-C_{28}	C_{10}-C_{29}, C_{31}	C_8-C_{30}
Protosalvinia	var.	var.	var.	var.
Prototaxites	C_{11}-C_{36}	C_7-C_{11}, C_{14}	C_{11}-C_{16}	C_{13}-C_{15}, C_{17}-C_{20}
Pseudosporochnus	C_8-C_{26}	C_7-C_{20}	C_9-C_{22}	C_9-C_{23}
Psilophyton cf. *forbesii*	C_9-C_{31}	C_9-C_{30}	C_9-C_{33}	C_8-C_{31}
Psilophyton princeps	C_9-C_{29}	C_9-C_{30}	C_9-C_{12}	C_{10}-C_{28}
Renalia	C_6, C_8-C_{26}, C_{28}	C_7-C_{18}	C_8-C_{21}	C_9-C_{26}
Rhynia	C_9-C_{35}	C_9-C_{28}	C_{10}-C_{29}, C_{31}	C_8-C_{26}
Sawdonia	C_8-C_{28}	C_7-C_{20}	C_{11}-C_{22}	C_9
Sciadophyton	C_8-C_{28}	C_9-C_{14}	C_{10}-C_{20}	C_{13}-C_{24}
Solenites	C_9-C_{33}	C_9-C_{333}	C_{11}-C_{32}	C_8-C_{28}, C_{30}-C_{31}
Spongiophyton	C_6, C_8-C_{24}	C_5-C_{14}, C_{16}	C_{13}-C_{17}	C_8-C_{20}
Taeniocrada	C_8-C_{28}	C_7-C_{14}	C_{12}-C_{20}	C_{13}-C_{24}
Tetraxylopteris	C_9-C_{33}	C_7-C_{31}	? C_{12}	--
Triphyllopteris	C_{10}-C_{33}	C_{11}-C_{30}	C_{10}-C_{33}	C_8-C_{28}, C_{30}
Zosterophyllum	C_8-C_{26}	C_7-C_{20}	C_9-C_{20}	C_9-C_{23}

Table 4.   (cont.)

Dicarboxylic acids ($C_nH_{2n-2}O_4$)	Aromatic dicarboxylic acids	Branched acids	Keto acids ($C_nH_{2n-2}O_3$)	Aromatic Tricyclic acids
? C_{11}-C_{25}	C_{12}-C_{20}	C_9-C_{22}	C_{12}-C_{13}	C_9-C_{21}
C_8, C_{10}-C_{16}	C_7, C_8, C_{11}-C_{18}	C_3, C_7-C_9	C_{14}-C_{18}, C_{20}-C_{22}	C_{13}-C_{20}
C_{10}-C_{19}	C_8-C_{22}	C_9, C_{11}-C_{19}, C_{22}	C_{11}-C_{17}	C_9-C_{21}, C_{23}-C_{27}
C_{11}-C_{26}	C_{12}-C_{21}	C_9-C_{22}	C_{12}-C_{13}	C_{10}-C_{20}
C_{10}-C_{15}	C_9-C_{16}	C_{10}, C_{12}-C_{15}	C_{10}	C_{10}-C_{23}
C_{11}-C_{14}	C_{10}-C_{18}	C_{11}-C_{16}	C_9-C_{16}, C_{18}	--
C_{11}-C_{13}	C_{10}-C_{18}	C_{10}-C_{12}	C_{10}-C_{15}	C_{16}
C_{12}-C_{14}	C_{10}-C_{16}	C_{10}, C_{12}	C_{10}, C_{12}-C_{14}	C_{10}-C_{22}
C_{12}-C_{14}	C_{10}-C_{16}	C_{10}, C_{12}	C_{10}, C_{12}-C_{14}	C_{10}-C_{22}
C_{11}-C_{14}	C_{10}-C_{18}	C_{14}-C_{16}, C_{10}-C_{14}	C_{10}-C_{15}	C_{24}-C_{25}
C_{11}-C_{13}	C_{10}-C_{18}	C_{10}-C_{14}	C_{10}-C_{15}	--
C_{11}-C_{13}	C_{10}-C_{17}	C_{10}-C_{15}	C_{10}-C_{15}	C_{16}, C_{19}
C_4-C_{10}, C_{12},C_{16}	C_7-C_{14}	C_8-C_{11}	C_{10}-C_{18}	C_{10}
C_{11}-C_{25}, C_{27}	C_8-C_{22}	C_4, C_{11}-C_{19}, C_{22}	C_{15}	C_9-C_{21}
C_{12}-C_{11}	C_9-C_{14}	C_8-C_{20}	C_8-C_{15}	C_{10}-C_{14}
C_8-C_{17}	C_7-C_{18}	C_5	C_{13}-C_{20}	--
C_4C_{16}	C_7-C_{11}	C_5, C_8	C_{14}-C_{22}	C_{15}-C_{20}
C_{11}-C_{21}	C_9-C_{19}	C_{10}, C_{12}-C_{19}	C_{10}-C_{16}	C_9-C_{22}
var.	var.	var.	var.	var.
C_4-C_{10}	C_7-C_{16}	--	--	--
C_{11}-C_{16}	C_{10}-C_{18}	C_{10}-C_{14}	C_{15}	C_9-C_{21}
C_{11}-C_{16}	C_9-C_{17}	C_{10}, C_{12}-C_{21}	C_9-C_{15}	C_{10}-C_{24}
C_{11}-C_{16}	C_9-C_{17}	C_{10}, C_{12}-C_{19}	C_{10}-C_{15}	C_{10}-C_{25}
C_{11}-C_{15}	C_9-C_{16}	C_{10}, C_{12}, C_{14}-C_{19}	C_{10}-C_{15}	C_9-C_{21}
C_{11}-C_{21}, C_{23}	C_9-C_{19}	C_{10}, C_{12}-C_{19}, C_{21}	C_{14}	C_9-C_{23}
C_{11}-C_{14}, C_{16}	C_{10}-C_{14}	C_{12}	C_9-C_{18}	C_{10}-C_{22}
C_4-C_{12}, C_{14}	C_7-C_9, C_{11}-C_{17}	C_{10}	C_{10}	C_{11}-C_{20}
C_9-C_{25}	C_{11}-C_{21}	C_{10}-C_{25}	C_{12}-C_{15}	C_9-C_{21}
C_{12}-C_{14}	C_9-C_{15}, C_{18}	C_{10}-C_{14}	C_{10}-C_{14}	C_{12}-C_{13}
C_4-C_{12}, C_{14}	C_7, C_{11}-C_{17}	C_{10}	C_{10}	C_{11}-C_{20}
--	?C_{11}	--	--	C_9-C_{21}
C_{11}-C_{25}	C_{12}-C_{20}	C_9-C_{22}	C_{12}-C_{15}	C_9-C_{21}
C_{11}-C_{14}	C_9-C_{20}	C_9-C_{18}	--	--

Table 5. Classes of compounds used to assess taxonomic
          status of fossil plants.

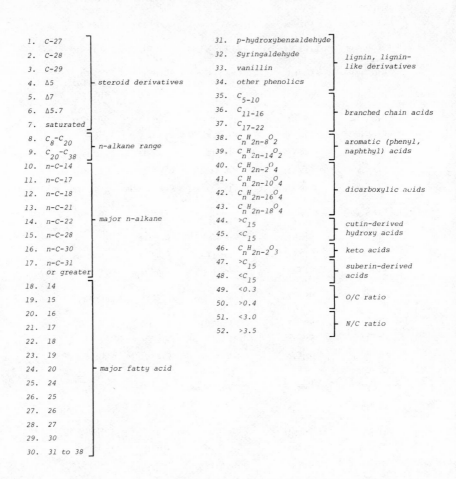

1.	C-27	⎤
2.	C-28	
3.	C-29	
4.	$\Delta 5$	steroid derivatives
5.	$\Delta 7$	
6.	$\Delta 5.7$	
7.	saturated	⎦
8.	C_8-C_{20}	⎤ n-alkane range
9.	C_{20}-C_{38}	⎦
10.	n-C-14	⎤
11.	n-C-17	
12.	n-C-18	
13.	n-C-21	
14.	n-C-22	major n-alkane
15.	n-C-28	
16.	n-C-30	
17.	n-C-31 or greater	⎦
18.	14	⎤
19.	15	
20.	16	
21.	17	
22.	18	
23.	19	
24.	20	major fatty acid
25.	24	
26.	25	
27.	26	
28.	27	
29.	30	
30.	31 to 38	⎦

31.	p-hydroxybenzaldehyde	⎤
32.	Syringaldehyde	lignin, lignin-like derivatives
33.	vanillin	
34.	other phenolics	⎦
35.	C_{5-10}	⎤
36.	C_{11-16}	branched chain acids
37.	C_{17-22}	⎦
38.	$C_nH_{2n-8}O_2$	⎤ aromatic (phenyl, naphthyl) acids
39.	$C_nH_{2n-14}O_2$	⎦
40.	$C_nH_{2n-2}O_4$	⎤
41.	$C_nH_{2n-10}O_4$	dicarboxylic acids
42.	$C_nH_{2n-16}O_4$	
43.	$C_nH_{2n-18}O_4$	⎦
44.	$>C_{15}$	⎤ cutin-derived hydroxy acids
45.	$<C_{15}$	⎦
46.	$C_nH_{2n-2}O_3$	⎤ keto acids
47.	$>C_{15}$	⎤ suberin-derived acids
48.	$<C_{15}$	⎦
49.	<0.3	⎤ O/C ratio
50.	>0.4	⎦
51.	<3.0	⎤ N/C ratio
52.	>3.5	⎦

show chemical and morphologic features suggestive of adaptation to desiccation (*e.g.*, cutinized spores, hydroxy acid-rich ergastic components). Cluster I separates from the others at the 65% level of similarity. Cluster II contains *Botryococcus,* a known green algal genus ranging from the Carboniferous to the present. By inferrence, *Parka* and *Pachytheca,* which cluster with *Botryococcus,* are chemically most similar to the green algae. Clusters III-VI represent, for the most part, plant fossils that are presumed or known (by virtue of having tracheids) to be vascular plant fossils. The vascular nature of some fossils in clusters III-VI is, however, clearly problematic, *e.g.*, *Eoshostimella* segregates into Cluster IV along with vascular (*Rhynia, Cooksonia*) and presumed vascular plant remains (*Taeniocrada, Eogaspesiea*). On the basis of the fossils segregated in Clusters III-VI, these groupings appear to be assignable, predominantly, to the progymnosperms (III), rhyniophytes (IV), trimerophytes (V), and zosterophyllophytes (VI).

Figure 4 is a principle component analysis of 32 plant fossils and represents the statistical similarities among these specimens based upon the character states of their total chemical compositions (table 5). The axes of this graph are dimensionless. Seven clusters or groupings of plant fossils are discernable by means of this technique. Cluster I corresponds to known or presumed green algae. Cluster II contains all the non-vascular plants broadly defined as non-green algae some of which are nematophytes (*Nematothallus,* N; *Prototaxites,* P$_x$). Cluster III contains plants that have been classified as rhyniophytes (*Rhynia,* Ry; *Cooksonia,* Co; *Taeniocrada,* Ta) and vascular plants whose taxonomic affinities are in doubt (*Renalia,* R). Cluster IV represents the remains of two presumed species of *Psilophyton* (Pf; Pp)

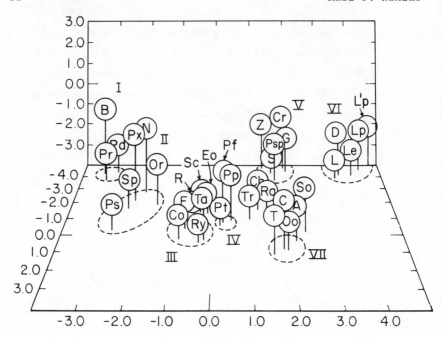

FIGURE 4. *Principle component analysis of 32 plant*
fossils (cf. Table 3) based on their chemical
compositions (cf. Table 4). Seven clusters
are discernable: presumed green algal taxa
(I), non-green algal taxa (II), rhyniophytes
(III), trimerophytes (IV), zosterophyllophytes
(V), lycopods (VI), and progymnosperms/
gymnosperms (VII).

a trimerophyte, and *Pertica* (Pt). Clusters V and VI contain
predominantly zosterophyllophyte and lycopod genera, respectively. Finally cluster VII contains plant fossils that are
known progymnosperms (*Archaeopteris,* A; *Tetraxylopteris,* T),
plants which may be precursors to the progymnosperms (*Oocampsa,*
Oo), and gymnosperm leaf cuticle (*Solenites,* So). Five fossils
do not cluster distinctly in any of the seven groupings:
Eohostimella, (Eo) and *Sciadophyton* (Sc), which fall near
the "rhyniophyte" cluster and *Chlidanophyton* (Ch),
Rhacophyton (Ra), and *Triphyllopteris* (Tr), which are flanked

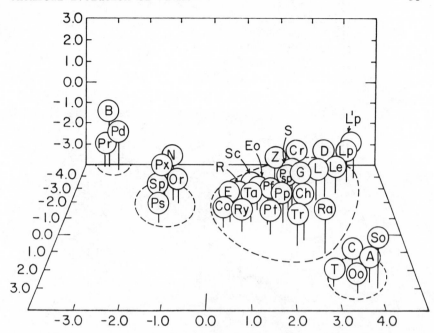

FIGURE 5. *Principle component analysis of taxa shown in Fig. 4, weighted to give emphasis to cutinic acid components.*

by the "trimerophyte", "progymnosperm" and "zosterophyllophyte" clusters.

Principle component analyses that have been weighted to give emphasis to cutinic acids (fig. 5), n-alkanes/fatty acids (fig. 6), and steranes (fig. 7), reveal varying numbers of clusters. The number and taxonomic composition of the clusters indicates the ability each of these chemical categories has to resolve differences among the suprageneric groups. When emphasis is given to cutinic acid compositions, four clusters are produced. Two of these correspond to non-vascular algal groups (Cluster I-II). The other two clusters correspond to "progymnosperms/gymnosperms" (Cluster IV), and a trimerophyte and lycopod complex (fig. 5). When priority is given to n-alkane/fatty acid compositions, the "lycopods"

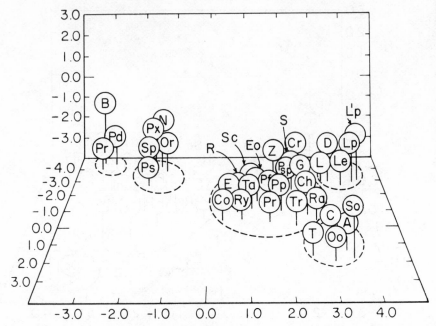

*FIGURE 6. Principle component analysis of taxa shown in
Fig. 4, weighted to give emphasis to n-alkane/
fatty acid components.*

are isolated from the vascular plant clustering in addition
to the "progymnosperm/gymnosperm" group (fig. 6). The best
resolution of the cluster configuration by means of weighted
data is achieved when priority is given to sterane constituents,
the diagenetic products of steroids (fig. 7).

From these analyses, numerous qualitative differences in
the chemical compositions of the fossils examined result in
statistically distinctive groupings. Clusters I and II of
fig. 4 are generally characterized by having short carbon
chain-length fatty acids/n-alkanes, lack extensive phenolic
derivatives, and show no evidence of true lignin biosynthesis.
In addition, the aromatic carboxylic acids isolated from these
fossils are generally less complex than those found in the
vascular plants. Cluster II is distinctive since it is

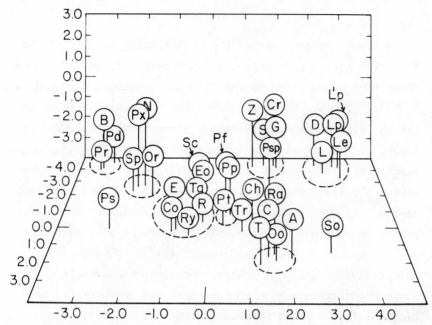

FIGURE 7. *Principle component analysis of taxa shown in Fig. 4 weighted to give emphasis to sterane components (=steroid derivatives).*

associated with a cutinic acid chemistry. Alkaline hydrolysis, transesterification and hydrogenolysis (with Li AlH$_4$ in tetra hydrofuran) of the remains of these fossils yields a mixture of monomers, and hydroxy and epoxy fatty acids (C$_{16}$ and C$_{18}$ constitutents, respectively). Clusters III–VII of fig. 4 reflect chemical compositions characterized by having lignin/ suberin derivatives, long carbon chain-length fatty acids/ alkanes, and complex aromatic carboxylic acids, steranes, phenols, and tannins. Qualitative differences in these constitutents, which are the cumulative effects when all the chemical constituents are considered, cause the distinct separation of the five "vascular" plant clusters.

Angiosperms

Attempts to reconstruct the evolutionary history of the
flowering plants have been based predominantly on inferences
drawn from comparative morphologic and biogeographic studies
of modern forms. While these studies have culminated in more
natural systems of classification (Cronquist, 1968; Takhtajan,
1969), and more realistic biogeographic models for angiosperm
distribution (Raven and Axelrod, 1974; Schuster, 1976), more
recent stratigraphic correlations among pollen and leaf fossils
have contributed much to the direct evaluation of angiosperm
evolution (Wolfe and Pakiser, 1971; Wolfe *et al.*, 1975; Doyle
and Hickey, 1976; Hickey and Doyle, 1977). Paralleling
comparative morphologic analyses of living and fossil
angiosperms, interest in the possible use of chemistry and
biochemistry in plant systematics has provided adjunctive
information to understanding flowering plant classification
and evolution (Alston and Turner, 1963; Swain, T., 1966;
Harborne, 1970; Luckner *et al.*, 1976; Bell and Charlwood,
1980). While organic geochemical analyses of fossil
angiosperms are currently limited to a few localities of
Eocene and Miocene age and to preliminary studies of some
Early Cretaceous fossils, chemical comparisons between fossil
and modern taxa have the potential to provide some of the most
detailed, direct evidence for biochemical evolutionary events
in the fossil plant record (cf. Niklas and Giannasi, 1977,
1978).

Currently, the oldest acceptable records for angiosperm
remains are monosulcate pollen grains from Barremian strata
of England, equatorial Africa, and the Potomac Group of North
America, and small, pinnately veined leaves with reticulate
venation from the basal Potomac Group and the Neocomian of
Siberia. Angiosperm leaves from the basal Potomac Group

make up only a very small component of the total flora found
in what Hickey and Doyle (1977) designate as Zone 1. These
leaves are disorganized in their venation, and show poor
differentiation of the petiole from the blade (Hickey and
Doyle, 1977). Chemical analyses of Zone 1 fossil leaves
reveal few biochemical components useful in resolving taxonomic
ranks below the level of class. On the basis of tannins,
lignin and cuticular biochemistry, leaves found in Zone 1
appear to have a generalized chemical profile consistent with
an assignment to the dicotyledons. Esters of glucose or
polyols with gallic, m-digallic, and hexahydroxydiphenic acids
or their congeners are restricted to the dicotyledons (Bate-
Smith, 1974). Hydrolysis of these acids yields the sugar core
and the constituent acid. Typical structures include tannic
acid, corilagin (1-galloyl-3,6- hexahydroxydiphenoylglucose),
and chebulagic acid. Analyses of Zone 1 fossils (unpublished
data) indicate these hydrolyzable tannins are present in
significant concentrations. Similarly, the lignin of most
monocotyledons have a lower proportion of sinapyl units in
comparison to dicotyledons. No such bias could be detected
in the Zone 1 fossil material. Significantly then, the
earliest recognizable angiosperm leaves have an already
canalized biochemical profile that is not only distinct from
gymnosperms but from other angiosperm groups. Finally, both
quantitative (e.g., the extent of cutinic acid oxidation) and
qualitative differences (cutin acid composition) found in the
cuticular composition of the Zone 1 material are compatible
with a dicotyledonous origin. This last line of evidence
must be used with great caution, however, since significantly
different cutin compositions in leaves may occur even during
their development and an adequate survey of angiosperm
cutinic acid variability is not available. Unfortunately,

the scarcity of the Zone 1 fossils from the Potomac Group
of North America and the extent of apparent leaching that has
occurred in the relatively course grain strata bearing these
fossils, currently do not permit a detailed chemotaxonomic/
paleobiochemical analysis. Based on the data that are
available, these fossils appear to have phytochemical affinities
with the Chloranthaceae (Piperales). The bases for these
similarities are tenuous at best, however, owing to the lack
of a complete survey of modern forms and of the need for a
more complete study of the fossils.

Chemical analyses of younger angiosperm material has
yielded a greater diversity of compounds that are more useful
in the phytochemical and chemotaxonomic characterizations.
Analyses of Lower Miocene angiosperm compression fossils from
the Succor Creek of Oregon, show the presence of pheophorbides
and carotenoids. In addition, flavonoids have been isolated
from these fossils and provide a series of compounds capable
of species diagnosis (Giannasi and Niklas, 1977). To date,
five angiosperm leaf taxa have been chemically placed in direct
apposition to modern forms thought to be their putative
descendents (table 6). These data have provided an opportunity
to discuss the extent of biochemical alterations that have
occurred within a lineage during the last 20 million years
(Niklas and Giannasi, 1978). For example, *Acer oregonianum*
is morphologically similar to the extant *A. macrophyllum*
found in Western America (table 6). The fossil taxon
characteristically possesses a flavonoid profile consisting
primarily of flavonols (*e.g.*, quercetin and kaempferol-3-o-
glycosides) and tannins. The tannins found in the fossil
Acer occur consistently in a number of extant Asian species
rather than in Western American maple species. Biochemically,
the fossils are most similar to the extant *A. carpinifolium*

Table 6. Floristic affinities of fossil angiosperm taxa based upon morphological and paleobio-chemical data.

| Fossil species | Morphologically similar species | | | Paleobiochemical affinity | |
	E.America	E.Asia	W.America	Geography	Taxonomy
Acer bendirei	A.saccharinum				
Acer columbianum			A.glabrum		
Acer glabroides	A.rubrum				
Acer minor	A.negundo		A.negundo		
Acer oregonianum*		A.henryi	A.macrophyllum	E.Asia	A.carpinifolium
Acer scottiae		A.pictum group**			
Celtis ?obliquifolia*	C.occidentalis complex			E.Amer.	C.occidentalis
Quercus consimilis*		Q.stenophylla: Q.myrsinaefolia		E.Asia	Q.acutissima: Q.chenii
Quercus dayana	Q.virginana				
Quercus eoprinus	Q.montana				
Quercus hannibali			Q.chrysolepis		
Ulmus newberryi	U.americana				
Ulmus paucidendata*	U.alata			E.Amer.	U.alata
Ulmus speciosa	U.racemosa				
Zelkova browni (=Z. "oregoniana")			Z.serrata	E.Asia	Z.serrata

*Taxa preserved in ash-fall deposits.

**Acer pictum group is now considered part of the section Platanoidea.

of eastern Asia. Similarly, leaves of *Celtis,* the hackberry,
are chemically similar to *C. occidentalis* which is found
currently in eastern North America. The flavonoid analysis
of fossil *Celtis* and *Ulmus* species demonstrated an absolute
dichotomy between flavonoid classes found in these two genera.
Fossil *Celtis* species possess only glycoflavones, while the
fossil *Ulmus* possess only flavonols, specifically quercetin
3-0-glycosides. This dichotomy has since been shown to be a
reliable intergeneric distinguishing feature and is supported
by the exclusive presence of glycoflavones in living *Celtis*
(*e.g., C. occidentalis*). Likewise, flavonols are the
dominant flavonoids in extant *Ulmus* species (Santamour, 1972;
Bate-Smith and Richens, 1973; Giannasi, 1978a). As an out-
growth of the biochemical studies of fossil representatives
of the elm family, such as *Ulmus* and *Celtis,* Giannasi (1978a)
has demonstrated flavonoid based distinctions on the sub-
familial level in both fossil and living genera. Similar
analyses of fossil angiosperms found in Miocene deposits of
the St. Maries River (Clarkia) area of Northern Idaho (cf.
Smiley *et al.,* 1975), have yielded diagnostic chemical
profiles useful at the generic and familial levels.

While the flavonoids are extremely distinctive compounds
by which fossil species may be distinguished and compared to
living taxa, cycloalkanes are more frequently encountered in
fossil angiosperm material and are useful in determining
ordinal or familial levels. The geologically occurring
steranes have their biological counterparts in the steroids.
Stereochemical transformations may be used to reconstruct the
nature of the steroid from which each sterane was derived.
Table 7 lists the currently known distributions of cycloalkanes in representative orders of plants (unpublished data).
A number of dicotyledon orders show the same cycloalkane

Table 7. *Paleochemotaxonomic distributions of cycloalkane*
distributions in representative orders of plant
fossils.

Order	*Cycloalkane distribution*
Sapindales	*Oleanane*
Caryophyllales	*Oleanane*
Myrtales	*Oleanane*
	Lupane
Hamamelidales	*Oleanane*
	Lupane
Rubiales	*Oleanane*
	Ursane
Rosales	*Oleanane*
	Lupane
Geraniales	*Oleanane*
	Ursane
	Lupane
Fagales	*Oleanane*
	3,4-seco-Oleanane
	Ursane
	Lupane
	Glutane (D:B-friedo-O)
	Friedelane (D:A-friedo-O)
Celastrales	*Friedelane*
Rhamnales	*Lupane*
	abeo-Lupane
Myricales	*Taraxerane*
Umbellales	*Ursane*
Rutales	*Ursane*
	Arborane
Leguminales	*Oleanane*
	Onocerane
	Ergostane
	Stigmastane
Coniferales	*Hopane*
	Ursane
Filicales	*Fernane (E:C-friedo-Hopane)*
	Hopane
	Acliantane (30-hor-Hopane)
Sphagnidae	*Ursane*
	Taraxerane
	β-Sitostane

distribution (*e.g.*, oleanane and lupane are found in representatives of the Myrtales, Hamamelidales, Rosales, Geraniales,
and Fagales), however, taken in tandem with other chemical
constituents, the organic profiles for these orders are
diagnostically different. Cycloalkanes, in addition to
serving as useful taxonomic marks, are extremely significant
in that they provide some clues to the evolutionary events
associated with steroid biochemistry within specific lineages
(see for example Nes *et al.*, 1977).

SECONDARY METABOLITE BIOSYNTHESIS: ROUTES AND MECHANISMS

   The dividing line between primary and secondary metabolism
is an indistinct one. There are many rare amino acids that
are definitely secondary metabolites, while many steroid
alcohols play essential structural roles in most organisms,
and may be considered as primary metabolites. There are three
principal starting materials for secondary metabolism (fig. 8):
(1) shikimic acid, which is the precursor of many aromatic
compounds, (2) amino acids, leading to alkaloids in addition
to proteins, and (3) acetate, the precursor of polyphenols
and the isoprenoids such as terpenes, steroids, and carotenoids.
The elaboration or merging at various levels within the
primary (and later the secondary) biosynthetic hierarchy of
metabolites has resulted in the evolutionary appearance of
compounds novel for a particular organism or lineage. These
new metabolites have to be maintained in the biochemical
profiles of various taxa for various lengths of evolutionary
time under the appropriate selective pressures.

   There are at least two biosynthetic methods by which
novel metabolites can be produced: (1) the elaboration of a
basic skeletal type by the extension of a single metabolic

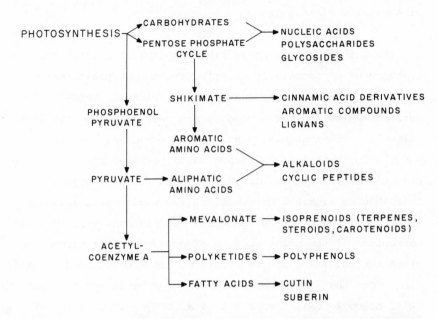

FIGURE 8.   *Simplified flow diagram illustrating the three*
            principal starting materials and some of the
            resulting products for secondary metabolism.

pathway, and (2) the merging of two pre-existing pathways.
In both cases, modification of the pre-existing enzymatic
system(s) is required.   Lignin biosynthesis may be considered
as an example of a novel metabolite (polymer) derived from a
single biosynthetic system by means of elaboration.   The
biosynthesis of numerous alkaloids may serve as examples of
merging between two pathways (*e.g.*, indole and benzylisoquinoline alkaloides).   The resultant metabolite from either
of these two methods may subsequently form the precursor
for a "second generation" of metabolite via biosynthetic
elaboration or as a contributing component of a merging
pathway.

Biosynthetic Elaboration

The unique aspects of lignin biosynthesis are (1) the presence of an ammonia-lyase effective on phenylalanine or tyrosine, and (2) the hydroxylation of cinnamic acid-derivatives to the corresponding p-coumaric acid derivatives (fig. 9). Other requisite enzymes (peroxidases, esterases, phenolases, methyl-transferases and esterified carboxyl group reductases) for lignin biosynthesis are found in organisms incapable of lignin biosynthesis. The presence of ammonia-lyases would have had an immediate and profound affect on the metabolism of the plant capable of producing this enzyme, since the resulting active esters of ring-substituted cinnamic acids form the biosynthetic starting compounds for flavonoids (*via*, malonyl CoA), coumarines (*via*, o-hydroxylation), various glycosides, esters of glucose quinate, hydroxybenzoic acids, and tannins (fig. 9). Subsequent elaboration of the cinnamic acid derivatives from the phenylpropanoid acids of plants has led to a variety of polymers called lignins, dependent upon the relative concentrations of the cinnamic acid derivatives available. For example, conifer lignin is particularly rich in coniferyl alcohols, while dicot lignins are rich in sinapyl alcohol.

The biosynthesis of cutin and suberin may have also involved only a single innovative biosynthetic step incorporating fatty acids (fig. 8). The formation of the dihydroxylated C_{16} and C_{18} acid monomers, which represent the major components of cutin and suberin, requires an initial w-hydroxylation of the corresponding palmitic and oleic, linoleic, or linolenic acids. Subsequent hydroxylation at C-10 and C-9 involves pre-existing dehydration mechanisms. The conversion of w-hydroxy fatty acids to carboxylic acids is the unique reaction involved in the biosynthesis of the

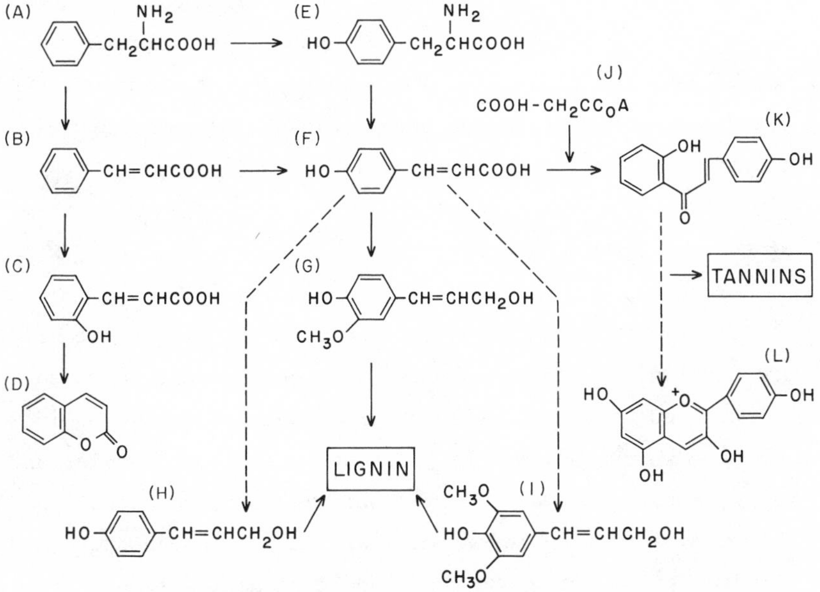

FIGURE 9. Simplified biosynthetic pathways associated
with the formation of lignin and tannins:
A. phenylalanine, B. cinnamic acid, C. o-
coumaric acid, D. coumarin, E. tyrosine,
F. p-coumaric acid, G. coniferyl alcohol,
H. p-coumaryl alcohol, I. sinapyl alcohol,
J. malonyl-CoA, K. chalcones, L. antho-
cyanidins.

the major aliphatic components of suberin. Polymerization of

cutinic acids to form cutin itself may be at least partly a

nonenzymatic process (Kolattukudy and Walton, 1973).

The appearance of a complete biosynthetic pathway for

plant alkaloids probably requires moderate genetic alterations.

These alterations can take the form of a shift in metabolite

compartmentalization, a slight decrease in enzyme specificity,

or the production of a particularly high concentration of an

amino acid (Mothes, 1969; Rosenberg and Stohs, 1974). There

are very few types of unprecedented reactions in alkaloid

biogenesis, since the mechanisms of condensation, oxidation, and reduction that are important in the transformation of amino acids to alkaloids have many parallels in primary metabolism. Robinson (1955) has even produced many of these reactions nonenzymatically under mild pH and temperatures, and low precursor concentrations.

Biosynthetic Merging

Classic examples of a compound of mixed biogenetic origin are provided by the metabolite mycophenolic acid (derived from the acetate and mevalonate pathways), the isoprenoid quinones which include the plastoquinones, tocophenols, and menaquinones (derived from the shikimate and mevalonate pathways), and the flavonoids (derived from the acetate and shikimate pathways). Complex indole alkaloids appear to be the result of amino acid and monoterpenoid metabolism (Battersby, 1971; Bisset, 1975) involving the condensation of the tryptamine amino group with the secologanin aldehyde group. Similarly, it is likely that the benzylisoquinoline alkaloids are the result of the merging of two branches of tyrosine metabolism -- the condensation of dopamine and 3,4-dihydroxyphenylpyruvate (Geissman and Crout, 1969).

The merging of two biogenetic pathways may result in a rapid phytochemical radiation in diversity. The flavonoids, and indole and benzylisoquinoline alkaloids are produced in a bewildering array of variants. Similarly, these compounds may be the basis for subsequent elaboration, *e.g.*, benzylisoquinoline is modified to yield benzyophenanthridine, protoberberine, and the aporphine alkaloids. Qualitative variation of biosynthetically related secondary compounds within natural or cultivated plant populations is well

documented for alkaloids (Dolinger *et al.*, 1973; Tabata
and Hiraoka, 1976), monoterpenes (Thorin and Nommik, 1974),
terpenoids (von Rudloff, 1975), and sesquiterpene lactones
(Mabry, 1970). Where such variation has been examined
genetically, the number of loci involved has been small.

Mechanisms for Biogenetic Changes

The mechanisms by which the synthesis of novel metabolites
are made possible include (1) structural-gene mutations, which
produce enzymes with dramatically altered substrate specificity,
(2) regulatory-gene mutations, which yield enzymes with relaxed
substrate specificity, (3) loss of a catabolic or anabolic
step resulting in the accumulation of metabolites that normally
undergo rapid conversion, (4) enzyme complementation brought
about by hybridization and (5) polymorphic variation in the
primary structure of enzymes (=isozymes). Based on microbial
enzyme studies, the production of a pre-existing enzyme in high
concentration that shows a low but appropriate level of activity,
is far more likely than structural-gene mutations. This may
be followed by secondary structural gene mutations that effect
the active site of the enzyme so that its reactivity increases.
Structural-gene mutations, such as gene duplication, may have
the same effect as regulatory-gene mutations which increase the
concentration of an enzyme with relaxed substrate specificity.
Such mutations would be advantageous if selective pressures
favored the retention of both the old and new function of the
enzyme. Selection for specificity-modifying mutations within
one of the duplicate genes could be envisaged as the first
step in the dual enzyme system, followed by a series of
changes that would restore the regulation of the unmodified
structural-gene and impose a new system for regulating the

modified structural gene.

Beside structural- or regulatory-gene mutations, several lines of evidence suggest that relatively simple genetic changes can result in dramatic biochemical changes. Certain novel flavonoids found in polyploid species have been shown to be the result of the accumulation of biosynthetic inter- mediates of end products common to the parental species (*e.g.*, Levy and Levin, 1971, 1974). Enzyme complementation in hybrids has also been shown to produce new compounds (Levy and Levin, 1975). Similarly, the nearly universal occurrence of protein polymorphism may play a substantial role in producing secondary metabolite variations in natural populations (cf. Selander, 1978).

ADAPTATIONS AND SELECTION PRESSURES FOR SECONDARY METABOLITES

While it is possible to uncover the biosynthetic mechanisms by which such critical metabolites as lignin, suberin, cutin, alkaloids, and steroids may have come into existence, the nature of the selective regimes that made these biogenetic elaborations advantageous must remain speculative. Even a brief survey of the phytochemical literature will indicate that sweeping generalizations concerning the function of some of these compounds is impossible, *e.g.*, steroids function as structural components of cell membranes, and as allelochemics (glycosidic steroid alkaloids); tannins may serve as UV screens, structural components in cell walls, or as repellants. Some paleobiochemical data are, however, useful in proposing and perhaps testing some hypotheses.

Lignin

Perhaps the most evolutionary significant secondary
metabolite in plants is lignin. Some of the functions
ascribed to lignin are: (1) regulators of the hydration of
hydrophilic moieties in cell walls, (2) bulking agents capable
of resisting concertina compression in cell walls, and (3)
as a protective agent against pathogenic attack and consumption
by herbivores (Friend, 1976). Lignin precursors and phenolics
associated with satellite biogenetic pathways (*e.g.*, hydroxybenzoic acid derived phenols, p-hydroxycinnamyl alcohol derived
lignins and phenylmopenes, and p-hydroxycinnamic acid derived
flavonoids and xanthones) are known to behave as plant growth
regulators: (1) lunularic acid is a dormin-like hormone in
liverworts (Pryce, 1972), fulfilling the function of abscisic
acid in vascular plants, (2) flavonoids with a catechol B-ring
have a sparing effect on auxin-oxidase activity, (3) flavonoids
with a monohydroxy phenol B-ring augument auxin-oxidase
activity, as do various hydroxycinnamic acids, and (4) some
p-coumaric acid esters are necessary co-factors for ethylene
biosynthysis (cf., Mapson, 1970), while caffeic acid inhibits
ethylene production. The production of phenolics would have
given a selective advantage to the first plants producing
them by conferring the capacity to produce potential growth
regulators, allelochemicals, and intermediate metabolites
from which flavonoids, courmarins, various glycosides, and
tannins could be synthesized. On the other hand, the activity
of many of these compounds can be detrimental to the plant
producing them. Various phenolics react synergistically with
hormones and are capable of inhibiting ATP synthesis,
uncoupling respiration, or interferring with ion absorption
(Stenlid, 1970). Plants capable of producing phenolics

detoxify these compounds by sequestering them by subcellular
compartmentalization and/or biosynthetic elaboration involving
polymerization. Phenolics produced by algae are excreted in
large quantities (Sieburth and Jensen, 1969). Extracellularly,
these compounds serve as ultra-violet absorbing agents (Yentsch
and Reichert, 1962) and growth inhibitors to bacteria, fungi
and other algae (Craigie and McLachlan, 1964). In addition,
some phenolic substances liberated by algae appear to be
necessary for their normal development and the completion of
life cycles, *e.g.*, the green algae *Ulva* and *Monostroma*
(Provasoli, 1965). Thus detoxification by excretion into an
aqueous or semi-aquaeous environment has circumvented the
detrimental effects of phenolic biosynthesis, while the
microbial pathogenicity of these compounds have obvious
benefits. Since flavonoids are found in some algae, particularly
the charophytes, the additional selective advantage of the
relevant, biogenetic pathway of these UV screening compounds can
be inferred.

  The excretion of phenolics and other metabolites becomes
limited by surface area to volume relationships for which
water loss considerations must take at least equal priority
in terrestrial habitats. The localization of phenolic
constituents in cuticular-like, ergastic layers over the
surfaces of the plant or in vacuoles is a non-enzymatic
process involving little energy and could have provided
immediate benefits: (1) an effective generalized defense
mechanism against predation, (2) a UV shield against
photooxidation, and (3) a molecular mechanism to regular
external water flow over the plant surface similar to the
ectohydric system of bryophytes. Polymerization of phenolics
into larger molecules would have also been an efficient
detoxification mechanism, and would provide various benefits,

e.g., structural rigidity for self support, a feeding deterrent, and a molecular way of selectively hydrolyzing the walls
of cells specialized to conduct water. T. Swain and Cooper-
Driver (1981) have suggested that phenolics in cuticles and
cell walls may have been the enucleation sites for the non-
enzymatic polymerization of lignin-like moieties.

From comparative studies it is evident that biochemical,
anatomical, and life history canalization of the land plants
occurred rapidly. In particular, a distinct dichotomy is seen
between the mosses and liverworts, and the tracheophytes. The
former lack lignin and tracheids, do not produce endogenous
auxin (Sheldrake, 1971), and have their gametophyte as the
dominant generation. The tracheophytes, in addition to lignin
and auxin biosynthesis, have the sporophyte as their dominant
life form. This dichotomy may have been the result of genetic,
hence biosynthetic, limits to each group, and/or the result
of different selection pressures involving life history
strategies. Both the bryophytes and tracheophytes are thought
to be derived from an ancestral green algal plexus that had
the capacity to produce chlorophylls a and b, particular
carotenoids, starch, sporopollenin, cellulose and hydroxy-
proline-rich cell walls, together with cytologic features
such as a specific chloroplast and flagellar infrastructure,
and cell division by means of a phragmoblast. In addition,
Stafford (1974; see also Loffelhardt *et al.*, 1973) has shown
that all green plants capable of producing ubiquinones also
have the capacity to produce phenylalanine ammonia-lyase,
while Siegel and Siegel (1970) have shown that the green
algae possess a peroxidase with a similar specificity to
that found in higher plants capable of bringing about lignin
synthesis. If the bryophytes and tracheophytes shared the
same ancestral lineage, then on the basis of shared primitive

characters the bryophytes most probably had the capacity to
produce trans-cinnamate from phenylalanine, but did not
elaborate on this compound to produce lignin. Sarkanen and
Hergert (1971) consider the polyphenols produced by various
mosses to be polyflavonoids with ferulic ester groups. Since
these compounds have typically a low methoxyl content, Sarkanen
and Hergert speculate that a genetic inability to methylate
p-coumarate derivatives may account for the lack of lignification seen in the bryophytes. In vascular plants, the
enzymatic reduction of ring-substituted cinnamic acids to
their corresponding alcohols appears to involve three enzymes
(cinnamate: CoA ligase, cinnamoyl-CoA reductase, and
cinnamyl alcohol dehydrogenase), while the dehydrogenative
polymerization of the primary lignin precursors involves
hydrogen peroxide, peroxidases, and free-radical formation
(Stafford, 1974; Gross, 1976). Since some of the dehydrogenative enzymes necessary to polymerize lignin from cinnamyl
alcohols have been reported for the mosses (e.g., coniferyl
alcohol dehydrogenase, cf. Mansell et al., 1976), the absence
of lignin in the bryophytes may be due to the lack of a
coordinated enzymatic system.

Lignification, auxin biosynthesis, and xylem differentiation
are involved in a complex interactive system. In green plants,
the presence of auxin (indol-3yl-acetic acid) as an endogenously
produced hormone correlates with lignin formation. Sheldrake
(1973) suggests that cells undergoing autolysis can be a major
source of auxin. Differentiating vascular tissue and
lignification is known to relate to auxin gradients. Lewis
(1980), therefore, proposed a causal relationship between the
accumulation of phenolics in boron deficient cells, and a
release or synthesis of auxin due to the autolysis of cells.
While there appears to be a correlation, the interpretation of
a direct causal relationship between auxin formation and

lignification is mitigated by the fact that (1) lignification
of xylem cells occurs late in their development and well after
the cells have passed through their maximum auxin content,
and (2) auxin appears to inhibit peroxidase activity. The
oxidation of auxin catalyzed by lAA-oxidase, a strong
peroxidase, during subsequent stages of xylem differentiation
may provide a more direct link between lignification and
gradients of auxin and autolytic cells. Seen in this
perspective, the metabolic catabolism of auxin in the pro-
vascular tissue may be correlated with peroxidase activity
and the dehydrogenative polymerization of alcohols into lignin.
Since bryophytes do not produce endogenous auxin, the establishment of a peroxidase countergradient, which would define a
pattern of lignification, does not occur.

While perhaps diagnostic of inherent genetic differences,
the biochemical dichotomy observed between the bryophytes and
tracheophytes may also be directly related to inferred selective
pressures referable to life history characteristics. The
"structural polyflavonoids" isolated from mosses are biochemically similar to lignin in that they are capable of
(1) regulating the hydration of cell walls, (2) selectively
defining a pattern of cell wall hydrolysis, thus permitting
a directional preference to water flow in specialized cells
(cf. Scheirer, 1980, p. 302), and (3) having allelochemic
activity against bacteria and fungi. In terms of functional
attributes, moss polyphenols are, therefore, identical to
lignin but for their limited ability to resist compressional
failure. An evaluation of the selective pressures operative
on plants which produce their motile male and non-motile female
gametes on separate axes suggests that there must be inherent
limits to the extent of vertical growth, which if exceeded
would severely reduce the efficiency of sexual reproduction

unless specialized dispersal mechanisms evolve (Niklas *et al.*,
1980). The production of tracheids and lignin would not be
favored in a life history where the gametophyte generation
is dominant, since both are unnecessary, and if present,
could reduce reproductive efficiency. The production of
highly efficient water conducting cells such as tracheids and
of a physiologic and architecturally versatile polymer such
as lignin would, however, be advantageous in plants where the
dominant generation is the sporophyte.

Tracheids, Hemitracheids, and Tubes

The morphologic and biochemical characteristics of some
cell types found in Silurian and Lower Devonian strata provide
evidence that the complete canalization between the bryophytes and tracheophytes may not have occurred until the
Upper Devonian. Cells that are morphologically identical
to tracheids have been reported from Upper Silurian *Cooksonia*
(Edwards and Davies, 1976). However, chemical analyses of
similar material reveals a lignin moiety that is similar to
that found in early Silurian "banded tubes" (Niklas and
Pratt, 1980) that have been interpreted to be the remains of
non-vascular plants (Pratt *et al.*, 1978). In turn, banded
tube cell types have structural characteristics similar to
both tracheids and the water conducting cells of some mosses
(=hydroids). Elongated cell types lacking internal ribbing or
secondary thickening have been isolated from a variety of
plant fossils (*e.g.*, nematophytes) while tubular cells have
been seen in specimens of plants traditionally classified as
tracheophytes (*e.g.*, *Taeniocrada, Rhynia major*). The
characteristics of modern tracheids, hydroids, and "tubes"
isolated from vascular plants, liverworts (*e.g.*, Metzgeriales),

marine algae (*e.g.*, "trumpet cells" of the Phaeophyta), and
"nematophytes" are summarized in fig. 10. These are compared
to those features preserved in Silurian "banded tubes".
Three criteria are used here to distinguish tracheids from
non-tracheid cell types; structure, biochemistry, and inferred
function. All three criteria are considered essential, since
there exists considerable variability in each, even for modern
tracheids. Tracheids may be defined as specialized cells
functioning to conduct water, elongated with tapered or inclined
end walls, non-living at maturity, and possessing lignified
secondary walls with thickenings and a potential variety of
pit types (cf. Bierhorst, 1960; Esau, 1977). Functionally
mature elongated xylem elements, which are dead at maturity,
have been reported to lack secondary thickenings (List, 1958),
while the extent of lignification (virtually none in *Equisetum)*
and its distribution in the compound middle lamella and
secondary wall vary among plant groups (cf., Mark, 1967).
While lignin has been reported in hydroids (Siegel, 1969),
a number of researchers have failed to show lignin in mosses
(Hébant, 1977; Miksche and Yasuda, 1978). Modern hydroids
are extremely variable in appearance (cf. Hébant, 1974, 1977)
and often show considerable similarity with tracheids.

On the basis of biochemical and developmental studies,
the canalization of tracheids versus other cell types
adapted to water conduction may have followed the outline
given in fig. 11. An ancestral plexus of green algae is
thought to have given rise to a group (or groups) of plants
capable of surviving on land. One structural requisite
envisioned was the formation of elongated cells that could
preferrentially conduct water. Characteristic features
of these tubular cells were dictated by the physiologic
and morphologic constraints attending the apoplastic transport

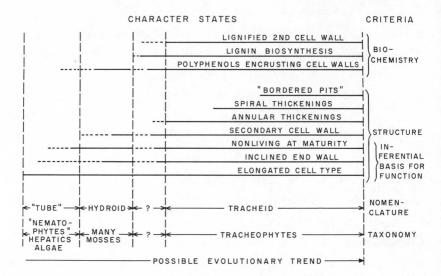

FIGURE 10. *Summary of the character states and criteria used to identify tracheids and hydroids from other cell types (e.g., "tubes") found in Silurian-Devonian strata. Cell types isolated from the Silurian (Llandoverian shown by ? have intermediate characteristics between tracheids and hydroids.*

of water, these include preventing cavitation within the water column and cell wall implosion due to the negative pressure induced by water flow. Canalization of biosynthetic pathways alternatively favoring polyphenols or lignins occurred in plants which had the gemetophyte and sporophyte as their dominant generation, respectively. The term "hemitracheid" is used here to designate the generalized water conducting cell type of the earliest land plants which subsequently gave rise to the hydroids in bryophytes and the tracheids in tracheophytes. As such, it refers to a level of organization, not to a taxonomic group. The inferrential characteristics of this intermediate cell type are summarized in fig. 10. With

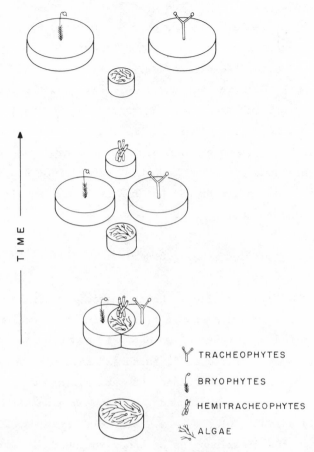

FIGURE 11.   *Simplified, hypothetical scenario depicting
             the canalization of bryophytes and tracheo-
             phytes from an antecedent group ("hemitracheo-
             phytes") which had its origin in an ancestral
             plexus of green algae. The term "hemi-
             tracheid" is used to describe the elongated
             cell type that has the shared primitive
             characters of both tracheids and hydroids
             (assuming that both bryophytes and tracheo-
             phytes share a common lineage). Other
             lineages not referable to the green algae
             or peripheral to the green algal lineages
             successful on land which may have evolved
             elongated "tube" cells are not shown.*

progressive elaboration of the biosynthetic pathways, and
morphology, three large groups of plant defined by their
level of organization are envisioned to exist in the Upper
Silurian-Devonian; the bryophytes, tracheophytes, and the
relicts of the hemitracheophytes. Clearly, not all tubular
cell types isolated from Silurian strata are referable to
the "hemitracheophytes"; evidence is accruing that lineages
derived from non-green algal lines may have temporarily
invaded the land. The "nematophytes" may reflect a number of
taxa some of which may or may not be direct ancestral forms
to early land plants.

Secondary Metabolites as Allelochemics

With the advent of lignin biosynthesis, branching
patterns could evolve, capable of resisting compression
stresses imposed by vertical weight. In addition to the
various monomers related to "structural" secondary metabolites,
a vast array of other compounds that are not directly referable
to cuticle, suberin, or lignin formation have been isolated
from early vascular plants, (cf. table 4). Statistical
analyses of the distributions of these compounds in major
plant lineages, show significant intertaxonomic differences.
The data indicate that shortly after the appearance of the
first vascular land plants, a rapid biochemical diversification
occurred, involving in some cases potentially fundamental
biogenetic alterations. The composition of the "lignin"
components extracted from the early rhyniophytes and other
paleobiochemically similar plants are distinctly different
from those isolated from the younger trimerophytes, lycopsids,
or progymnosperms (cf. fig. 4). By contrast the trimerophytes
and progymnosperms are quantitatively very similar with regard

to their lignin composition. The zosterophyllophytes and
lycopsids are different from other groups with regard to
their phenols and branched chain acids. Tannins, in the
broad sense, are often the most abundant constituents
isolated from Devonian vascular fossils. These compounds are
bound in protein-polysaccharide complexes (melanoidines) and
are insoluble copolymers at normal pH. Extraction of
melanoidins with hot 80% aqueous methanol yields the phenolic
components. In most cases, these phenols have a similar
skeleton to that of catechin, a flavan-3-ol, and to that
found in β-tannins. Phenols of this molecular type, however,
are lacking in fossils having a "zosterophyllophyte" paleobiochemistry, but are found in relatively high concentrations
in fossil lycopods. While the paleobiochemical compositions
of the early land plants parallel some distinctive morphologic
features for each lineage, indicating that both biochemical
and morphologic features were rapidly canalized, the
selection pressures that led to this canalization are
currently unknown. Kevan *et al.* (1975) argue that terrestrial
arthropods and fungi may be responsible for various puncture
lesions and necrotic areas found on the axes of Devonian
vascular plant fossils, and that various outgrowths (enations)
on axial surfaces, and possibly arborescence may be the result
of predation. T. Swain and Cooper-Driver (1981) have suggested
that some of the phenolic compounds isolated from fossil plants
may be remnants of toxins and feeding deterrents. Many phenols
isolated from modern vascular plants appear to limit the
amount of material ingested by a herbivore in a dosage-
dependent fashion (cf. Friend and Threlfall, 1976; Chapman
and Bernays, 1978; Friend, 1979). Such quantitative defense
substances are often present in high concentrations, up to
60% dry weight in the case of tannins. Janzen has suggested

that polyphenol-rich plant associations were probably
"... populated by very few herbivores and the plants in
them were very well defended with chemical traits..." (Janzen,
1979, p. 339). In modern ecosystems, tannins (1) form
relatively indigestible complexes with proteins thereby
reducing the rate of dietary nitrogen assimilation (cf. Feeney,
1969), (2) inhibit a variety of microbial and animal digestive
enzymes (Goldstein and Swain, 1965), (3) reduce the rate of
bacterial protein hydrolyses (Handley, 1961), and (4) inhibit
fungal growth. Similarly lignins reduce the digestibility of
plant carbohydrates and proteins by hydrogen-bonded complex
formation, in an analogous way to that of plant tannins. In
addition, these compounds are structural macromolecules that
make plant tissues mechanically indigestible. Thus lignins
and tannins may constitute the two most significant classes
of quantitative defensive substances in plants. The presence
of quantitative toxins and mechanical repellants in early
vascular land plants is consistent with the suggestion by
Rhoades (1979) that such deterrents are common in plants that
are predictable food resources. The selection pressure of the
high metabolic costs of generalized chemical defense systems
may account for the apparently rapid biochemical diversification
seen in Paleozoic plants (Rhoades and Cates, 1976; Feeny, 1976;
Chew and Rodman, 1979).

Gymnosperms

Very little is known, from direct paleobiochemical
comparisons, about the biochemical diversification of seed
plants during the Late Paleozoic and Mesozoic. Attempts to
determine the rates of non-flowering seed plant evolution
indicate that the gymnosperms have evolved slowly in
comparison with angiosperms, *e.g.*, genera of gymnosperms are

about as old as orders of angiosperms. Prager *et al.*, (1976)
have concluded that the rate of change in the amino acid
sequence of seed proteins between two lineages is about 1%
per 7.5 million years. They speculate that the slow rate of
anatomical evolution within the Pinaceae is possibly related
to their slow rates of chromosomal evolution. Based on mean
species duration times, Levin and Wilson (1976) estimated that
the net speciation rates of conifers and cycads are 0.02 and
0.01, respectively, in comparison to those of angiosperm herbs
and shrubs, 1.05 and 0.24 respectively. These authors propose
that differences in the breeding structures of the species
studied and in environmental predictability may account for
the significantly slow rates of karyotypic and organismal
evolution seen in the non-flowering seed plants. Coincidental
with these observations is the generalized defense posture of
most conifers and the long induction and relaxation times
(1-5 years) during which allelochemicals respond to predation
pressure (cf. Rhoades, 1979; table 1). By inference,
relatively slow rates of biochemical diversification would
be expected among the gymnosperm lineages. While the vicariant
distributions of some secondary metabolites among gymnosperm
lineages preclude any generalization, Langhammer *et al.*,
(1977) report a polysaccharide gumtar component from the
leaves of *Welwitschia* that is found elsewhere only in the
Magnoliaceae, Mimosaceae, Apiaceae, Meliaceae, Rutaceae and
the gymnosperm *Araucaria bidwillii*). It does appear that
biochemical diversification among the gymnosperms has not
progressed as far as it has in the more recently evolved
angiosperms. For example, the gymnosperms typically produce
flavonols including myricetin, C-glycoflavones, and biflavonyls.
In contrast, angiosperms produce all classes of flavonoids and
demonstrate complex glycosylation and extra oxygenation steps

in the biosynthesis of these compounds (cf. Swain, T., 1975).
Care must be taken, however, in the use of these data since
it is often difficult to distinguish between primitive and
advanced conditions in flavonoid phenotypes (cf. Gornall
and Bohm, 1978).

Angiosperms

   With the appearance of angiosperms in the Cretaceous, the
elaboration and diversification of many classes of secondary
metabolites appear to have reached a zenith. Many of these
compounds function as cues to pollinators, seed dispersal
agents, and as toxins, reflecting the coevolution of the
flowering plants and animals. The often disjunct distribution
of predator repellents (tannins, iridoids, polyacetylenes,
sesquiterpene lactones) and attractants (flavonoids) within
angiosperm groups suggests that these taxa experienced ecologic
and physiologic reiterations of stress in their history. When
the chemical arsenal of an angiosperm group was breached by a
predator, new repellants evolved allowing their possessors
to enter a new adaptive zone and diversify. A long period of
disuse may permit the revival of an old set of repellants
(cf. Cronquist, 1977). The reproductive necessity to attract
pollinators, on the one hand, and to repell pests on the other
hand, has led to a compartmentalization of secondary metabolites
in various plant parts which while precedented in gymnosperms
is unparalleled in any other plant group. Perennial parts
such as woody stems generally contain quantitative feeding
repellants that are immobile and non-reclaimable (lignin,
tannins) while actively growing parts often contain small
amounts of qualitative repellants (alkaloids, cyanogenic
glycosides, glucosinolates). Floral parts often show the
most complex spacial and temporal allocations of specific

metabolites.

While the morphology and biochemical profiles of the few
Miocene angiosperm fossils studied have distinct modern
counterparts at the species level, some compounds isolated
from these fossils (steroid derivatives) appear to be
biosynthetically simpler (primitive?) then those found in
modern taxa. Interestingly, if the paleobiochemical and
morphologic data are taken in tandem, no evidence of stasis
for these taxa has been uncovered. The isolation of flavonoids
from Miocene angiosperm remains has provided the most sensitive
method whereby fossil plants may be directly compared with
living taxa. The biosynthetic pathway and relevant genetics
of flavonoids are understood in detail (Hahlbrock and Grisebach,
1975; Giannasi, 1978b; Wong, 1976) and when correlated with the
presence of distinct flavonoids in fossils may provide a basis
for inferring relative rates of genetic change in specific
angiosperm lineages.

Yet another avenue of current research is the correlations
between biochemical profiles and detailed ultrastructural
isolation of organic components. Electron microscopy of
Miocene angiosperm tissues on occasion reveal well preserved
chloroplast infrastructure and other organelles (fig. 12).
Cytochemical investigations of these tissues may provide
information on the intracellular compartmentalization of
secondary metabolites and on the diagenesis attending
fossilization (Niklas *et al.*, 1978). Current interest in
the evolution of angiosperms has centered on until recently
comparative studies among modern forms resulting in a large
body of morphologic, biochemical, and taxonomic data. When
placed within this detailed framework, biochemical and
ultrastructural studies of fossil angiosperms may give us
the most comprehensive understanding of the evolution of a
specific plant group.

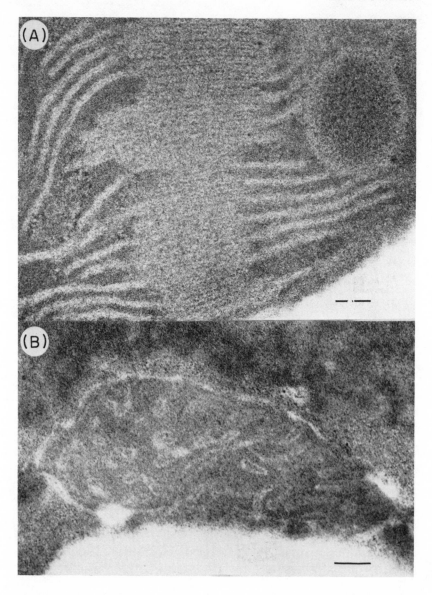

LITERATURE CITED

ALSTON, R.E., and B.L. TURNER. 1963. Biochemical Systematics. New Jersey: Prentice-Hall.

BANKS, H.P. 1968. The early history of land plants. *In:* Evolution and Environment. New Haven: Yale University Press, Pp. 73-107.

BANKS, H.P. 1975. The oldest vascular land plants: a note of caution. Rev. Palaeobot. Palynol. 20:13-25.

BATE-SMITH, E.C. 1974. Systematic distribution of ellagitannins in relation to the phylogeny and classification of the angiosperms. *In:* Chemistry in Botanical Classification. New York: Academic Press. Pp. 93-102.

BATE-SMITH, E.C., and R.H. RICHENS. 1973. Flavonoid chemistry and taxonomy of *Ulmus.* Biochemical Systematics and Ecology 1:141-146.

BATTERSBY, A.R. 1971. Biosynthesis-II: Terpenoid indole alkaloids. *In:* The Alkaloids. Specialist reports, Vol. 1. London: Chemical Society. Pp. 31-47.

BELL, E.A., and B.V. CHARLWOOD, (eds). 1980. Secondary Plant Products. Berlin: Springer-Verlag.

BIERHORST, D.W. 1960. Observations on tracheary elements. Phytomorphology. 10:249-305.

BISSET, N.G. 1975. Chemical structures and biosynthesis of Loganiaceae alkaloids. Pharm. Weekbl. 110:425-441.

CHALONER, W.G., and A. SHEERIN. 1979. Devonian microfloras. The Devonian system. Special papers in Paleontology. 23:145-161.

CHAPMAN, R.F., and E.A. BERNAYS, (eds.). 1978. Insect and Host Plant. Ent. Exp. & Appl. 24:201-766.

CHEW, F.S., and J.E. RODMAN. 1979. Plant resources for chemical defense. *In:* Herbivores: Their Interaction with Secondary Plant Metabolites. New York: Academic Press. Pp. 271-307.

FIGURE 12. Transmission electron micrographs of fossil angiosperm leaf tissues from the Miocene Clarkia locality. The bars in both micrographs represent 100nm. A. Portion of a chloroplast from fossil Betula *(a birch) showing the grana fretwork membrane system. B. Mitochondrion and protoplasmic remnants from a fossil* Hydrangia *floral bract. Reprinted with permission, from BioScience, February 1981. ©American Institute of Biological Sciences.*

CRAIGIE, J.S., and J. McLACHLAN. 1964. Excretion of coloured
   ultraviolet absorbing substances in marine algae. Can.
   J. Bot. 42:23-33.
CRONQUIST, A. 1968. The Evolution and Classification of
   Flowering Plants. Boston: Houghton Mifflin.
CRONQUIST, A. 1977. On the taxonomic significance of
   secondary metabolites in angiosperms. Plant Syst. Evol.
   Suppl. 1:179-189.
DOLINGER, P.M., P.R. EHRLICH, W.L. FITCH, and D.E. BREEDLOVE.
   1973. Alkaloid and predation patterns in Colorado lupine
   populations. Oecologia   13:191-204.
DON. A.W.R., and G. HICKLING. 1917. On *Parka decipiens*.
   Q.J. Geol. Soc. Lond. 71:648-666.
DOYLE, J.A., and L.J. HICKEY. 1976. Pollen and leaves from
   the mid-Cretaceous Potomac Group and their bearing on
   early angiosperm evolution. *In:* Origin and Early Evolution
   of Angiosperms. New York: Columbia University Press.
   Pp. 139-206.
EDWARDS, D., and E.C.W. DAVIES. 1976. Oldest recorded *in situ*
   tracheids. Nature 263:494-495.
EDWARDS, D., M.G. BASSETT and E.C.W. ROGERSON. 1979. The
   earliest vascular land plants: continuing the search for
   proof. Lethaia 12:313-324.
ESAU, K. 1977. Anatomy of Seed Plants, 2nd Edition. New York:
   John Wiley and Sons.
FEENEY, P.P. 1969. Inhibitory effect of oak leaf tannins on
   the hydrolysis of proteins by trypsin. Phytochemistry
   8:2119-2126.
FEENEY, P.P. 1976. Plant apparency and chemical defense.
   Recent Adv. Phytochem. 10:1-40.
FRIEND, J. 1976. Lignification in infected tissue. *In:*
   Biochemical Aspects of Plant Parasite Relationships.
   New York: Academic Press. Pp. 291-304.
FRIEND, J. 1979. Phenolic substances and plant disease.
   Rec. Adv. Phytochem. 12:557-588.
FRIEND, J., and D.R. THRELFALL. 1976. Biochemical Aspects of
   Plant Parasite Relationships. London: Academic Press.
GEISSMAN, T.A., and D.H.G. CROUT. 1969. Organic Chemistry of
   Secondary Plant Metabolism. San Francisco: Freeman,
   Cooper, Inc.
GIANNASI, D.E. 1978a. Generic relationships in the Ulmaceae
   based on flavonoid chemistry. Taxon. 27(4):331-344.
GIANNASI, D.E. 1978b. Systematic aspects of flavonoid
   biosynthesis and evolution. Bot. Rev. 44:339-429.
GIANNASI, D.E., and K.J. NIKLAS. 1977. Flavonoid and other
   chemical constituents of fossil Miocene *Celtis* and *Ulmus*
   (Succor Creek Flora). Science 197:765-767.

GOLDSTEIN, J.L., and T. SWAIN. 1965. Changes in tannins in ripening fruits. Phytochemistry 2:371-383.

GOOD, B.H., and R.L. CHAPMAN. 1978. The ultrastructure of *Phycopeltis* (Chroolepidaceae; Chlorophyta). I. Sporopollenin in the cell walls. Amer. J. Bot. 65:27-33.

GORNELL, R.J., and B.A. BOHM. 1978. Angiosperm flavonoid evolution: A reappraisal. Systematic Botany 3:353-368.

GROSS, G.G. 1976. Biosynthesis of lignin and related monomers. Phytochem. Society of North America, Symposium and Annual Meeting, U. of British Columbia, Abstract No. 4.

HAHLBROCK, K., and H. GRISEBACH. 1975. Biosynthesis of flavonoids. *In:* The Flavonoids. London: Chapman and Hall, Pp. 866-915.

HANDLEY, W.R.C. 1961. Further evidence for the importance of residual leaf protein complexes in litter decomposition and the supply of nitrogen for plant growth. Plant & Soil 15:37-73.

HARBORNE, J.B. (editor). 1970. Phytochemical Phylogeny. London: Academic Press.

HÉBANT, C. 1974. Studies on the development of the conducting tissue-system in the gametophytes of some Polytrichales II. Development and structure at maturity of the hydroids of the central strand. Jour. Hattori Bot. Lab. 38:565-607.

HÉBANT, C. 1977. The Conducting Tissues of Bryophytes. Hirschberg: A.R. Ganter Verlag.

HICKEY, L.J., and J.A. DOYLE. 1977. Early Cretaceous fossil evidence for angiosperm evolution. Bot. Rev. 43:3-104.

JANZEN, D.H. 1979. New horizons in the biology of plant defenses. *In:* Herbivores: Their Interaction with Secondary Plant Metabolites. New York: Academic Press. Pp. 331-350.

JOHNSON, J.H., and K. KONISHI. 1958. Studies of Devonian algae. Q. Colo. School Mines 53:1-114.

KEVAN, P.G., W.G. CHALONER, and D.B.O. SAVILE. 1975. Interrelationships of early terrestrial arthropods and plants. Palaeontology 18:391-418.

KLITZSCH, E., A. LEJAL-NICOL, and D. MASSA. 1973. Le Siluro-Devonian a Psilophytes et Lycophytes du bassin de Mourzonk (Libye). C.R. Acad. Sci. Paris 277:2465-2467.

KOLATTUKUDY, P.E., and T.J. WALTON. 1973. The biochemistry of plant cuticular lipids. Prog. Chem. Fats Other Lipids Part 3, 13:121-175.

LANGHAMMER, L., M. HORISBERGER, I. HORMAN, and A. STRAHM. 1977. Partial structure of a polysaccharide from leaves of *Welwitschia mirabilis*. Phytochemistry 16:1575-1577.

LEVIN, D.A., and A.C. WILSON. 1976. Rates of evolution in seed plants. Net increase in diversity of chromosome numbers and species numbers through time. Proc. Natl. Acad. Sci. USA 73:2086-2090.

LEVY, M., and D.A. LEVIN. 1971. The origin of novel flavonoids in *Phlox* allotetraploids. Proc. Natl. Acad. Sci. USA 68:1627-1630.

LEVY, M., and D.A. LEVIN. 1974. Novel flavonoids and
    reticulate evolution in the *Phlox pilosa-P. drummondii*
    complex. Amer. J. Bot. 61:156-167.
LEVY, M., and D.A. LEVIN. 1975. The novel flavonoid
    chemistry and phylogenetic origin of *Phlox floridiana*.
    Evolution 29:487-499.
LEWIS, D.H. 1980. Boron, lignification and the origin of
    vascular plants -- a unified hypothesis. New Phytol.
    84:209-229.
LIST, A., JR. 1958. The embryogeny and seedling development
    of *Gleditsia triacanthos* L. Master's Thesis. Cornell
    University.
LOFFELHARDT, W., B. LUDWIG, and H. KINDL. 1973. Thylakoid-
    gebindene L-Phenylalanin-Ammoniak-Lyase. Hoppe-Seylers
    Z. Physiol. Chem. 354:1006-1012.
LUCKNER, M., K. MOTHES, and L. NOVER. (eds.). 1976.
    Secondary Metabolism and Coevolution. Leipzig: Nova Acta
    Leopold. Suppl. No. 7.
MABRY, T.J. 1970. Intraspecific variation of sesquiterpene
    lactones in *Ambrosia* (Compositae). Applications to
    evolutionary problems at the populational level. *In:*
    Phytochemical Phylogeny. New York: Academic Press.
    Pp. 269-300.
MANSELL, R.L., E.R. BARBEL, and M.H. ZENK. 1976. Multiple
    forms and specificity of coniferyl alcohol dehydrogenase from
    cambial regions of higher plants. Phytochemistry 16:1849.
MAPSON, L.W. 1970. Biosynthesis of ethylene and the ripening
    of fruit. Endeavour 39:29-33.
MARK, R.E. 1967. Cell Wall Mechanics of Tracheids. New
    Haven: Yale Univ. Press.
MATIC, M. 1956. The chemistry of plant cuticles; a study
    of cutin from *Agave americana* L. Biochem. J. 63:168-176.
MIKSCHE, G.E., and S. YASUDA. 1978. Lignin of "giant" mosses
    and some related species. Phytochemistry 17:503-504.
MOTHES, K. 1969. Die alkaloide im Stoffwechsel dur Pflanze.
    Experimentia 25:225-240.
NES, W.R., K. KREVITZ, J. JOSEPH, W.D. NES, B. HARRIS, and
    G.F. GIBBONS. 1977. The phylogenetic distribution of
    sterols in tracheophytes. Lipids 12:511-527.
NIKLAS, K.J. 1979. An assessment of chemical features for
    the classification of plant fossils. Taxon 28:505-516.
NIKLAS, K.J. 1980. Paleobiochemical techniques and their
    applications to paleobotany. *In:* Progress in Phyto-
    chemistry Vol. 6:143-182. Oxford: Pergamon Press.
NIKLAS, K.J., and W.G. CHALONER. 1976. Chemotaxonomy of
    some problematic Palaeozoic plant fossils. Rev.
    Palaeobot. Palynol. 22:81-104.

NIKLAS, K.J., and GENSEL, P.G. 1976. Chemotaxonomy of some
  Paleozoic vascular plants. Part I: Chemical compositions
  and preliminary cluster analyses. Brittonia 28:353-378.
NIKLAS, K.J., and D.E. GIANNASI. 1977. Flavonoids and other
  chemical constituents of fossil Miocene *Zelkova* (Ulmaceae).
  Science 196:877-878.
NIKLAS, K.J., and D.E. GIANNASI. 1978. Angiosperm paleo-
  biochemistry of the Succor Creek Flora (Miocene) Oregon,
  U.S.A. Amer. J. Bot. 65:943-952.
NIKLAS, K.J., and L.M. PRATT. 1980. Evidence for lignin-like
  constituents in Early Silurian (Llandoverian) plant fossils.
  Science 209:396-397.
NIKLAS, K.J., R.M. BROWN, JR., R. SANTOS, and B. VIAN. 1978.
  Ultrastructure and cytochemistry of Miocene angiosperm
  leaf tissues. Proc. Natl. Acad. Sci. USA 75(7):3263-3267.
NIKLAS, K.J., B.H. TIFFNEY , and A.H. KNOLL. 1980. Apparent
  changes in the diversity of fossil plants: a preliminary
  assessment. *In:* Evolutionary Biology. New York: Plenum
  Publishing Corp. Pp. 1-89.
PRAGER, E.M., D.P. FOWLER, and A.C. WILSON. 1976. Rates of
  evolution in Conifers (Pinaceae). Evolution 30:637-649.
PRAGER, E.M., A.C. WILSON, J.M. LOWENSTEIN, and V.M. SARICH.
  1980. Mammoth albumin. Science 209:287-289.
PRATT, L.M., T.L. PHILLIPS, and J.M. DENNISON. 1978. Evidence
  of non-vascular land plants from the Early Silurian
  (Llandoverian) of Virginia, USA. Rev. Palaeobot. Palynol.
  25:121-149.
PROVASOLI, L. 1965. Nutritional aspects of algal growth.
  Proc. Can. Soc. Plt. Physiol. 6:126-127.
PRYCE, R.J. 1972. The occurrence of lunularic and abscisic
  acids in plants. Phytochemistry 11:1759-1762.
RAVEN, P.R., and D.I. AXELROD. 1974. Angiosperm biogeography
  and past continental movements. Ann. Missouri Bot. Gard.
  61:539-673.
RHOADES, D.F. 1979. Evolution of plant chemical defense
  against herbivores. *In:* Herbivores: Their Interaction
  with Secondary Plant Metabolites. New York: Academic
  Press. Pp. 331-350.
RHOADES, D.F., and R.G. CATES. 1976. Toward a general theory
  of plant antiherbivore chemistry. Recent Adv. Phytochem.
  10:168-213.
ROBINSON, R. 1955. The Structural Relations of Natural
  Products. Oxford: Oxford University Press.
ROSENBERG, H., and S.J. STOHS. 1974. The utilization of
  tyrosine for mescaline and protein biosynthesis in
  Lophophora williamsii. Phytochemistry 13:1861-1863.
SANTAMOUR, F.S., JR. 1972. Flavonoid distribution in *Ulmus*.
  Bull. Torrey Bot. Club 99:127-131.

SARKANEN, K.U., and H.L. HERGERT. 1971. Classification and
    distribution. *In:* Lignins: Occurrence, Formation,
    Structure and Reactions. New York: Wiley. Pp. 1-18.
SCHEIRER, D.C. 1980. Differentiation of bryophyte conducting
    tissues: structure and histochemistry. Bull. Torrey Bot.
    Club. 107:298-307.
SCHOPF, J.M., E. MENCHER, A.J. BOUCOT, and H.N. ANDREWS. 1966.
    Erect plants in the early Silurian of Maine. U.S. Geol.
    Surv. Prof. Pap. 550-D: D69-D75.
SCHUSTER, R.M. 1976. Plate tectonics and its bearing on the
    geographical origin and dispersal of angiosperms. *In:*
    Origin and Early Evolution of Angiosperms. New York:
    Columbia University Press. Pp. 48-138.
SELANDER, R.K. 1978. Genic variation in natural populations.
    In: Molecular Evolution. Sunderland: Sinauer Assoc., Inc.
    Pp. 21-45.
SHELDRAKE, A.R. 1971. The occurrence and significance of auxin
    in the substrate of bryophytes. New Phytologist 70:519-526.
SHELDRAKE, A.R. 1973. The production of hormones in higher
    plants. Biol. Rev. 48:509-542.
SIEBURTH, J. McN., and A. JENSEN. 1969. Studies on algal
    substances in the sea. II. The formation of Gelbstoff
    (humic material) by pheophyte exudates. J. Exp. Mar.
    Biol. Ecol. 3:275-289.
SIEGEL, B.Z., and SIEGEL, S.M. 1970. Anomalous substrate
    specificities among the algal peroxidases. Amer. J.
    Bot. 57:285-287.
SIEGEL, S.M. 1969. Evidence for the presence of lignin in
    moss gametophytes. Amer. J. Bot. 56:175-179.
SIGLEO, A.C. 1978. Organic geochemistry of silicified wood,
    Petrified Forest National Park, Arizona. Geochimica et
    Cosmochimica Acta 42:1397-1405.
SMILEY, C.J., J. GRAY, and L.M. HUGGINS. 1975. Preservation
    of Miocene fossils in unoxidized lake deposits, Clarkia,
    Idaho. J. Paleontology 49:833-844.
STAFFORD, H.A. 1974. The metabolism of aromatic compounds.
    Ann. Rev. Pl. Physiol. 25:459-486.
STENLID, G. 1970. Flavonoids as inhibitors of the formation
    of adenosine triphosphate in plant mitochondria.
    Phytochemistry 9:2251-2256.
STROTHER, P.K., and A. TRAVERSE. 1979. Plant microfossils
    from Llandoverian and Wenlockian rocks of Pennsylvania.
    Palynology 3:1-21.
SWAIN, F.M., J.M. BRATT, S. KIRKWOOD, and P. TOBBACK. 1969.
    Carbohydrate components of Paleozoic plants. *In:* Advances
    in Organic Geochemistry 1968. Oxford: Pergamon Press.
    Pp. 167-180.

SWAIN, T. (editor). 1966. Comparative Phytochemistry.
    London: Academic Press.
SWAIN, T. 1975. Evolution of flavonoid compounds. *In:*
    The Flavonoids. London: Chapman and Hall. Pp. 1096-1129.
SWAIN, T., and G. COOPER-DRIVER. 1981. Biochemical evolution
    in early land plants. *In:* Paleobotany, Paleoecology, and
    Evolution. New York: Praeger Press.
TABATA, M., and HIROAKA, N. 1976. Variation of alkaloid
    production in *Nicotiana rustica* callus cultures. Physiol.
    Plantarum 38:19-23.
TAKHTAJAN, A.L. 1969. Flowering Plants: Origin and Dispersal.
    Edinburgh: Oliver and Boyd.
THORIN, J., and H. NOMMIK. 1974. Monoterpene composition of
    cortical oleoresin from different clones of *Pinus sylvestris*.
    Phytochemistry 13:1879-1881.
von RUDLOFF, E. 1975. Chemosystematic studies of the volatile
    oils of *Juniperus horizontalis, J. scopulorum* and *J.
    virginiana*. Phytochemistry 14:1319-1329.
WOLFE, J.A., and H.M. PAKISER. 1971. Stratigraphic inter-
    pretations of some Cretaceous microfossil floras of the
    Middle Atlantic states. U.S. Geol. Surv. Prof. Pap.
    750-B: B35-B47.
WOLFE, J.A., J.A. DOYLE, and V.M. PAGE. 1975. The basis of
    angiosperm phylogeny. Paleobotany. Ann. Missouri Bot.
    Gard. 62:801-824.
WONG, E. 1976. Biosynthesis of flavonoids. *In:* Chemistry
    and Biochemistry of Plant Pigments, Vol. 1, 2nd ed. New
    York: Academic Press. Pp. 464-526.
YENTSCH, C.S., and C.A. REICHERT. 1962. The interrelationship
    between water-soluble yellow substances and chloroplastic
    pigments in marine algae. Botanica Mar. 3:65-74.

CARBON ISOTOPES AND THE EVOLUTION OF C_4 PHOTOSYNTHESIS AND CRASSULACEAN ACID METABOLISM

James A. Teeri

Department of Biology
and
Committee on Evolutionary Biology
University of Chicago
Chicago, Illinois

The relative abundances of ^{13}C and ^{12}C in green plant tissue, expressed as the $\delta^{13}C$ value, are determined to a large extent during the initial carboxylation of atmospheric CO_2. In those plants in which the initial carboxylation occurs via ribulosebisphosphate carboxylase the tissue $\delta^{13}C$ values are typically $-27°/_{oo}$. In contrast, in plants in which the initial carboxylase is phosphoenolpyruvate carboxylase the tissue $\delta^{13}C$ values are typically $-13°/_{oo}$. Surveys of biomass $\delta^{13}C$ values have shown that there is considerable variation in these values among the families of vascular plants. Species that utilize only C_3 photosynthesis have $\delta^{13}C$ values of ca. $-27°/_{oo}$. Species that utilize C_4 photosynthesis have $\delta^{13}C$ values of ca. $-13°/_{oo}$. In contrast, species that exhibit Crassulacean acid metabolism have a range of biomass $\delta^{13}C$ values from -13 to $-27°/_{oo}$. In many such species there appears to be little environmentally induced plasticity in the tissue $\delta^{13}C$ value. However, in some species the $\delta^{13}C$ value of an individual plant is subject to considerable modification by environmental variables. In all cases thus far studied, species with relatively positive $\delta^{13}C$ values, e.g., $-13°/_{oo}$, appear to be derived from taxa with more negative $\delta^{13}C$ values.

INTRODUCTION

    Over the past two decades it has become clear that there
are at least three types of photosynthetic carbon metabolism
among vascular plants: C_3 photosynthesis, C_4 photosynthesis,
and Crassulacean acid metabolism (CAM). Most species utilize
only one of these types. Species that exhibit C_3 photosynthesis utilize ribulose bisphosphate carboxylase (RuBP
carboxylase) in the presence of light to fix atmospheric CO_2.
In contrast, species that exhibit C_4 photosynthesis and most
species that exhibit CAM utilize phosphoenolpyruvate carboxylase (PEP carboxylase) to fix atmospheric CO_2. In C_4 plants
atmospheric CO_2 uptake via PEP carboxylase occurs in the light.
In CAM plants atmospheric CO_2 uptake via PEP carboxylase occurs
in the dark. However, as will be seen later, some CAM plants
can capture atmospheric CO_2 in the light by RuBP carboxylase
via C_3 photosynthesis. The photosynthetic cells and tissues
of C_4 and CAM plants differ from each other as well as from
C_3 plants in a number of biochemical and structural properties.
C_3 photosynthesis is by far the most common and widespread
type, occurring in nearly all vascular plant families. Both
C_4 photosynthesis and CAM are far less common, occurring in a
relatively small number of families. In all cases studied,
it appears that C_4 photosynthesis and CAM have evolved in
lineages derived from taxa that utilize only C_3 photosynthesis.
    The stable carbon isotope mass spectrometer has been an
important tool in studying the photosynthetic phenotypes of
vascular plants. There are substantial differences in stable
carbon isotope composition among plants having different types
of photosynthesis. The relative abundances of ^{13}C and ^{12}C
in plant tissue are to a large extent determined at the time
of initial carboxylation of atmospheric CO_2 during photosynthesis (Troughton, 1979). The relative abundances of these

two stable isotopes are normally expressed in parts per
thousand ($^{\circ}/_{\circ\circ}$), relative to a standard, on the $\delta^{13}C$ scale
where:

$$\delta^{13}C = [\frac{^{13}C/^{12}C \text{ sample} - {}^{13}C/^{12}C \text{ standard}}{^{13}C/^{12}C \text{ standard}}] \times 1000$$

On this scale, the carbon in the lower atmosphere which is
about one percent ^{13}C has a $\delta^{13}C$ value of *ca.* $-7^{\circ}/_{\circ\circ}$. C_4
plants have biomass $\delta^{13}C$ values of approximately $-13^{\circ}/_{\circ\circ}$.
C_3 plants have biomass $\delta^{13}C$ values of approximately $-27^{\circ}/_{\circ\circ}$.
There is now considerable evidence that RuBP carboxylase
causes the large amount of discrimination against ^{13}C in C_3
plants, resulting in the C_3 biomass $\delta^{13}C$ values of *ca.*
$-27^{\circ}/_{\circ\circ}$. This carboxylation occurs in an 'open system' in the
sense that the stomates are open and the ^{13}C that is dis-
criminated against can diffuse out of the photosynthetic
tissue into the external atmosphere. PEP carboxylase exhibits
a much smaller discrimination against ^{13}C, which accounts for
the more positive $\delta^{13}C$ values of *ca.* $-13^{\circ}/_{\circ\circ}$ found in the
biomass of C_4 plants and CAM plants that take up CO_2 only in
the dark.

The three types of photosynthesis are similar in that the
photosynthetic carbon reduction cycle (PCR cycle = Calvin
cycle) is in each type the pathway by which carbon is incorporated into hexoses in the presence of light. The primary
characteristics distinguishing the three types of photosynthesis are the reactions in which carbon participates before
entry into the PCR cycle. In a C_3 leaf atmospheric CO_2
diffuses into the mesophyll cell chloroplasts, where RuBP
carboxylase catalyzes the uptake of CO_2 into the PCR cycle.
In C_4 photosynthesis, atmospheric CO_2 diffuses into the
mesophyll cells, where PEP carboxylase catalyzes the incor-
poration of CO_2 into C_4 dicarboxylic acids. The C_4 acids are

then translocated into bundle sheath cells that surround the vascular tissue. In contrast to C_3 plants, the PCR cycle is not active in the mesophyll cells of C_4 plants; rather the PCR cycle is active in the bundle sheath cells. The C_4 acids are decarboxylated in the bundle sheath cells and the CO_2 enters the PCR cycle via RuBP carboxylase. The available evidence suggests that the bundle sheath PCR cycle is a 'closed system' in the sense that all, or nearly all, of the CO_2 released from the C_4 acids enters the PCR cycle. In addition, it appears that very little free CO_2 diffuses from the atmosphere into the bundle sheath cells; instead, nearly all of the carbon entering the PCR cycle is first converted into C_4 acids. Thus, in C_4 plants, the discrimination against ^{13}C that occurs in the mesophyll cells is not further altered in the PCR cycle, because all of the carbon captured by PEP carboxylase subsequently enters the PCR cycle.

Bender (1968, 1971) observed that C_3 and C_4 plants exhibited consistent differences in $\delta^{13}C$ values. Plants known to be C_4 exhibited $\delta^{13}C$ values near $-13°/_{oo}$, in contrast C_3 plants had $\delta^{13}C$ values near $-27°/_{oo}$. A large number of subsequent surveys of C_3 and C_4 plants have confirmed this correlation of $\delta^{13}C$ values with C_3 and C_4 photosynthesis. So far, no species have been conclusively demonstrated to be capable of switching from C_3 to C_4 photosynthesis, or vice versa, in mature leaves. In most cases it is a straight-forward matter to classify a plant as either C_3 or C_4. The first $\delta^{13}C$ value indicative of C_4 photosynthesis was published in 1953, over a decade before the discovery of the C_4 pathway. Craig (1953) surveyed the $\delta^{13}C$ values of wood and leaves of some 25 species of terrestrial plants to determine naturally occurring variations in carbon isotopic composition. Twenty-four of these species had $\delta^{13}C$ values that clustered tightly

around a mean value of $-26°/_{oo}$. These 24 species are now
known to be C_3 species. The one remaining species, an
unidentified grass from southwest Kansas, had a $\delta^{13}C$ value of
$-12.2°/_{oo}$. At that time C_4 photosynthesis had not yet been
described and it was suggested that the relatively positive
$\delta^{13}C$ value might be a result of contamination of the carbon
in the grass by carbon derived from limestone in the substrate.
However, it is now known that over 50 percent of the grass
species of southwest Kansas are C_4, so Craig's measurement
probably was made on a C_4 species.

In the case of CAM the situation is more complex. The
PCR cycle in CAM plants is active in the light and is located
in the chloroplasts of the photosynthetic cells. For a CAM
plant that only takes up atmospheric CO_2 in the dark, the
following sequence of events occurs. In the dark, when the
stomates are open, atmospheric CO_2 diffuses into the photosynthetic cells. There the CO_2 is combined via PEP carboxylase into oxaloacetic acid and then to malic acid. This
malic acid accumulates in the large vacuoles characteristic of
CAM photosynthetic cells. In the subsequent daylight period
the stomates close tightly and no further CO_2 is taken up from
the external atmosphere. The PCR cycle is activated. Malic
acid is transported from the vacuoles, decarboxylated, and
the resultant CO_2 enters the PCR cycle via RuBP carboxylase.
The available evidence suggests that in this type of CAM
plant all of the atmospheric CO_2 that is initially taken up
by PEP carboxylase enters the PCR cycle as is the case in C_4
plants. Thus in this type of CAM plant the discrimination
against ^{13}C in atmospheric CO_2 fixed at night via PEP
carboxylase is not further altered as the carbon is converted
into hexoses. The mean $\delta^{13}C$ value of tissues in CAM plants
with exclusively nocturnal CO_2 uptake is about $-13°/_{oo}$.

Based on biochemical and physiological criteria, species in
over a dozen plant families were known to have CAM when the
mass spectrometer came into use in photosynthesis studies.
Surveys of the $\delta^{13}C$ values of well-documented CAM plants revealed much greater variability than was found in either C_3
or C_4 plants. The $\delta^{13}C$ values of various CAM species spanned
the entire range from *ca.* -12°/$_{oo}$ to *ca.* -27°/$_{oo}$. As will be
discussed later, the $\delta^{13}C$ value is an integrated measure of
the history of carboxylation of atmospheric CO_2 that makes up
the structural carbon of a tissue. The interpretation has
been that among plants capable of CAM there is phenotypic
variability with regard to the carboxylation of atmospheric
CO_2.

THE DEFINITION OF CAM

There are many phenotypic characters that are highly
correlated with the presence of CAM. However, many of these
characters themselves vary among taxa in their degree of
correlation with CAM. A survey of the many phenotypes that
CAM plants have yields the following definition of the basic
characteristics necessary to consider a plant to have CAM:
The photosynthetic cells have the ability to fix CO_2 in the
dark via PEP carboxylase, forming malic acid which accumulates
in the vacuole. During the following light period the malic
acid is decarboxylated, and the CO_2 enters the PCR cycle in
the same cell.

The above definition specifically does not require that
dark uptake of atmospheric CO_2 actually occur for a plant to
have CAM. The reason for this is that there is increasing
evidence that a number of plants may utilize CAM, as defined
above, only to recycle respiratory-derived CO_2. Some of these

plants synthesize substantial amounts of malic acid in the dark without any detectable uptake of atmospheric CO_2. $\delta^{13}C$ values for biomass of these plants are those characteristic of C_3 photosynthesis. The proposed definition concerns only the pathways of CO_2 within the photosynthetic cells, regardless of whether the CO_2 source is metabolic or atmospheric. All postulated paths for CO_2 fixation in a CAM cell are presented in fig. 1. As will become evident, this definition accommodates all taxa classified as being CAM.

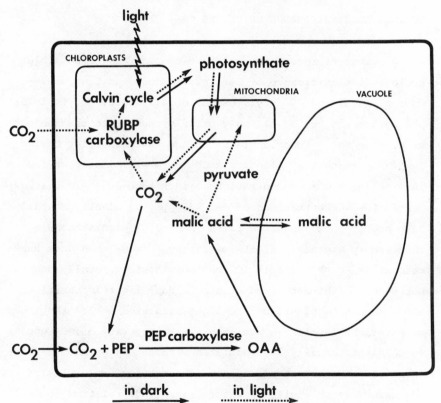

FIGURE 1. *Model of the pathways of CO_2 fixation theoretically possible in a CAM photosynthetic cell. The model is modified from Kluge (1976) and Kluge and Ting (1978).*

The distributions of C_4 photosynthesis and CAM are both sufficiently widespread that both types of metabolism are considered to have polyphyletic origins in vascular plants. This apparently repeated evolution of these two complex phenotypic characters has stimulated considerable debate about their modes of origin. The purpose of this chapter is to review evidence concerning evolutionary relationships of vascular plant groups in which C_4 photosynthesis and CAM occur.

ECOLOGICAL SIGNIFICANCE OF C_4 AND CAM

A number of recent authors have considered the possible ecological significance of CAM (*e.g.*, Kluge and Ting, 1978; Osmond, 1978) and C_4 (*e.g.*, Osmond *et al.*, 1980; Teeri, 1979) photosynthesis, so I will only briefly summarize the subject here. The major functional difference that separates C_4 and C_3 plants is that C_4 plants fix CO_2 more efficiently at low intercellular CO_2 concentrations. Thus even at ambient atmospheric CO_2 concentrations of *ca.* 335 ppm, C_4 plants generally have higher photosynthetic rates than do C_3 plants because PEP carboxylase has a greater affinity for CO_2 than does RuBP carboxylase. As a result, C_4 photosynthesis generally does not become light-saturated even at midday light intensities, whereas most C_3 plants become light saturated at 1/3 the midday light level. In addition the higher rate of CO_2 uptake in C_4 plants usually is associated with an approximate 2-fold increase in water use efficiency (the ratio of molecules of CO_2 gained to molecules of water transpired) relative to C_3 plants. Finally, most C_4 plants can maintain positive net photosynthetic rates at higher leaf temperatures than can most C_3 plants. Taken together, these attributes suggest that C_4 plants will have greatest competitive advantage in habitats

characterized by high temperatures, high levels of light and
low moisture. Analyses of the geographic patterns of
abundance of C_4 species (Teeri and Stowe, 1976; Stowe and
Teeri, 1978; Teeri, 1979; Teeri *et al.*, 1980) have shown that
in the families studied, C_4 species are generally most
abundant in climates with high temperatures. However, among
all C_4 species, there appear to be differences between
monocots and dicots in the particular type of high temperature
climate most favored. The distribution of C_4 monocots is
most highly correlated with growing season temperature (table
1). For example, similar proportions of the monocot floras
of both southern Florida and the Sonoran Desert are comprised
of C_4 species. In contrast, the distribution of C_4 dicots is
most highly correlated with growing season aridity (table 1).
In North America C_4 dicot species are most abundant in the
flora of the Sonoran Desert and adjacent desert regions. The
reasons for the different distributions of C_4 dicots and
monocots are not fully understood at present.

The major functional difference between CAM species and
non-CAM species is in the ability of CAM plants to fix CO_2 in
the dark and to accumulate it as malic acid in substantial
quantity in the photosynthetic cells. As a result, several
phenotypic differences between CAM and non-CAM plants are
possible. If all atmospheric CO_2 uptake in a CAM plant occurs
in the dark, then the water use efficiency of that plant can
be several-fold greater than that of a C_3 or a C_4 plant. The
reason for this is that at night the gradient in water vapor
pressure between the transpiring organ and the atmosphere is
much smaller than it is in the daylight, resulting in lower
transpiration rates at night. During periods of prolonged
drought, some CAM plants, at least, can keep their stomates
closed both day and night and still maintain low metabolic

Table 1. Environmental Variables Most Highly Correlated with the Distribution of C_4 Species in Five Plant Families and with the Distribution of Succulent Species in Two Plant Families

Family	r	Environmental Variable
C_4:		
Cyperaceae	+0.93	square of normal July minimum temperature
Gramineae	+0.97	normal July minimum temperature
Amaranthaceae	+0.85	summer pan evaporation
Chenopodiaceae	+0.83	annual dryness ratio
Euphorbiaceae	+0.88	summer pan evaporation
Succulent:		
Cactaceae	-0.93	log January coefficient of humidity
Crassulaceae	-0.74	July coefficient of humidity

rates by refixing respiratory CO_2 via dark accumulation of malic acid and subsequent fixation via the PCR cycle the following day. The plant cannot gain carbon during such a period, but it can balance energy losses and maintain a metabolic state that is capable of rapid resumption of growth when moisture once again becomes available. Finally, some CAM species have the ability to perform both daytime CO_2 fixation via C_3 photosynthesis and dark fixation of CO_2 into malic acid. Some of these plants appear to be capable of regulating their water use efficiency according to the changing moisture status of the environment by changing the proportions of day and night CO_2 fixation. In sum, the phenotypic attributes of CAM plants suggest they will be at their greatest competitive advantage in habitats characterized by low levels of moisture availability. We have analyzed the geographic patterns of

abundance of succulent species of the Cactaceae and Crassulaceae (Teeri *et al.*, 1978). All of these species are
thought to be capable of the accumulation of malic acid in
photosynthetic cells in the dark. In both families the
succulent species are most abundant in regions characterized
by dry soils. The single environmental variable most highly
correlated (negative correlation) with the abundance of these
species is the coefficient of humidity (table 1), an estimate
of soil moisture storage.

THE OCCURRENCE OF C_4 PHOTOSYNTHESIS

In vascular plants there is now strong evidence that C_4
photosynthesis occurs in at least 16 families (table 2;
Osmond *et al.*, 1980; Winter, 1981; Ziegler *et al* ., 1981).
The rate of discovery of new families in which C_4 occurs
has declined considerably in the past few years. However,
large regions of the earth, in particular tropical forests and
grasslands and the cool deserts of central Asia, may yield
additional C_4 families.

Thus far, C_4 photosynthesis has been found only in
flowering plants (table 2) with both monocots (2 families)
and dicots (14 families) having C_4 species. At present no
family has been found that contains only C_4 species.

The distribution of the C_4 families among the orders of
the flowering plants is presented in fig. 2 (modified from
Takhtajan, 1969). The most primitive dicot orders are in
the subclass Magnoliidae (Takhtajan, 1969; Cronquist, 1968)
and C_4 photosynthesis is not found in any of these orders.
It is currently thought that the subclass Magnoliidae
represents the basal complex out of which the remaining
angiosperms have been derived (Cronquist, 1968). The
orders of the subclass Hamamelidae also do not contain any

Table 2.   Vascular Plant Families with C_4 Photosynthesis

Anthophyta:
   Monocotyledoneae:
      Cyperaceae
      Gramineae

   Dicotyledoneae:
      Acanthaceae
      Aizoaceae
      Amaranthaceae
      Asteraceae
      Boraginaceae
      Capparidaceae
      Caryophyllaceae
      Chenopodiaceae
      Euphorbiaceae
      Nyctaginaceae
      Polygonaceae
      Portulacaceae
      Scrophulariaceae
      Zygophyllaceae

---

species with C_4 photosynthesis. This subclass appears to represent an early specialization in the evolution of flowering plants (Cronquist, 1968).

The remaining four dicot subclasses, the Caryophyllidae, Dilleniidae, Rosidae, and Asteridae, all contain some species with C_4 photosynthesis. Thus of the six subclasses of the dicots, C_4 photosynthesis is found in the four subclasses considered to be most highly derived (Cronquist, 1968).

Among the monocots there are four subclasses, only one of which, the Commelinidae, contains C_4 families (Cronquist, 1968). The phylogenetic relationships among the various monocot taxonomic groups are not resolved. However, the two C_4 families (Cyperaceae and Gramineae) appear to be closely related and to represent a derived condition among monocots (Cronquist, 1968; Clifford and Williams, 1980). Several fossil grasses from a Pliocene chert have been compared with regard to anatomy and $\delta^{13}C$ values (Nambudiri et al., 1978). Some of the fossils exhibit the mesophyll bundle sheath anatomy characteristic of extant C_4 grasses and these fossils had a $\delta^{13}C$ value of $-13°/_{\circ\circ}$. Other grass fossils in the same chert possessed the mesophyll anatomy characteristic of extant C_3 grasses and those fossils had a $\delta^{13}C$ value of $-24°/_{\circ\circ}$. Petrified roots in the chert are anatomically similar to extant species of *Paspalum* which is a C_4 genus.

Among the subclasses of flowering plants, no C_4 families occur in a subclass that is considered primitive.

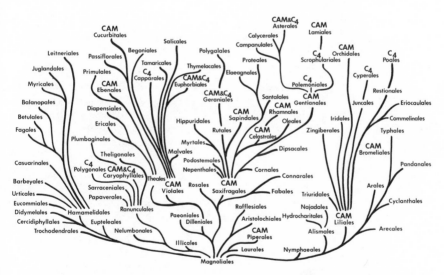

FIGURE 2.   The occurrence of C_4 photosynthesis and CAM
             among the orders of the Anthophyta. The
             dendogram follows Takhatajan (1969).

THE OCCURRENCE OF CAM

There is strong biochemical and physiological evidence for the presence of CAM in 26 families of vascular plants (table 3; Kluge and Ting, 1978; Rao *et al.*, 1979; Keeley, 1981). Preliminary morphological and physiological evidence suggests that some species of several other plant families may exhibit CAM. These families (Szarek, 1979) include: Bataceae, Carophyllaceae, Chenopodiaceae, Convolvulaceae, Passifloraceae, and Plantaginaceae and others. However, a complete biochemical

Table 3. Vascular Plant Families with CAM Photosynthesis

Lycophyta:
    Isoetaceae
Pteridophyta:
    Polypodiaceae
Gnetophyta:
    Welwitschiaceae
Anthophyta:

Monocotyledoneae:	*Dicotyledoneae:*
Agavaceae	*Aizoaceae*
Bromeliaceae	*Apocynaceae*
Liliaceae	*Asclepiadaceae*
Orchidaceae	*Asteraceae*
	Cactaceae
	Celastraceae
	Crassulaceae
	Cucurbitaceae
	Didiereaceae
	Ebenaceae
	Euphorbiaceae
	Flacourtiaceae
	Geraniaceae
	Labiatae
	Oxalidaceae
	Piperaceae
	Portulacaceae
	Sapindaceae
	Vitaceae

---

characterization of CAM has yet to be reported in any of these families. It appears likely that the list of CAM families, particularly in the angiosperms, will expand in the future. Habitats of these possible new CAM families include epiphyte habitats in tropical montane forests, cliff and bare rock surfaces in tropical and subtropical regions, coastal deserts, particularly fog deserts, and aquatic habitats.

Among the phylogenetic groupings of vascular plants CAM is more widespread than C_4 photosynthesis (table 3). CAM has been reported in the Isoetales *(Isoetes)*, Welwitschiales *(Welwitschia)* and the Filicales *(Drymoglossum* and *Pyrrosia)*. However, of the 26 CAM families, 23 are angiosperms including both monocots (4 families) and dicots (19 families).

Among the Lycophyta, the genus *Isoetes* appears to represent a derived condition (Taylor, 1981). Fossils similar to *Isoetes* are reported from Triassic and Cretaceous rocks and it is possible that the diminutive modern *Isoetes* may have been derived from an arborescent ancestor. At present there are no other reports of CAM in the Lycophyta; however, additional survey work is needed. As most species of *Isoetes* are primarily aquatic, the evolution of CAM in this group may have occurred as the Lycophyta invaded a new habitat, considerably differing in CO_2 availability from the terrestrial environment in which most other Lycophyte photosynthetic organs functioned.

Among the Filicales, the family Polypodiaceae contains two species which have been shown to have CAM. Both species are epiphytes and succulent (Kluge and Ting, 1978). Evolutionary relationships within the Filicales are uncertain; however, the oldest known fossils of the Polypodiaceae are from the Upper Cretaceous (Taylor, 1981). The family appears to represent at least a partially advanced condition within the Filicales. At present it appears that many ferns do not have CAM, but tropical epiphytic taxa clearly deserve additional study.

The Gnetales as a group are difficult to interpret and there is disagreement as to the relationships of extant taxa. CAM has been described in *Welwitschia* but has not yet been found in *Ephedra* or *Gnetum* (Kluge and Ting, 1978). However, these three genera are generally thought to be highly specialized among gymnosperms (Taylor, 1981).

The distribution of the CAM families among the orders of the flowering plants is presented in fig. 2. CAM is present in the most primitive subclass, Magnoliidae, in the order Piperales. Within the Magnoliidae, the Piperales occupy a derived position probably evolved from the Magnoliales (Cronquist, 1968). The orders of the Hamamelidae have not yet been reported to contain any CAM species.

CAM species occur in four families of the Caryophyllidae in the order Caryophyllales. The Caryophyllales are the most primitive order of the subclass (Cronquist, 1968). The four CAM families (Aizoaceae, Cactaceae, Didieraceae and Portulacaceae) are thought to be derived from the Phytolaccaceae which is the most primitive family of the Caryophyllidae. However, some members of the Phytolaccaceae can be interpreted as being succulent, and the presence of CAM may yet be described for this family.

CAM species occur in three families (Cucurbitaceae, Ebenaceae and Flacourtiaceae) of the Dilleniidae. Takhtajan (1969) places the Cucurbitaceae in the order Cucurbitales, but Cronquist (1968) includes the Cucurbitaceae in the Violales, in which the Flacourtiaceae is also placed. The Ebenaceae is in the order Ebenales. While the Flacourtiaceae is the most primitive family of the Violales, its possible

ancestral relationship to the advanced family Cucurbitaceae
is uncertain (Cronquist, 1968). Both the Ebenales and
Violales appear to be derived from the Theales.

CAM species occur in seven families (Celastraceae,
Crassulaceae, Euphorbiaceae, Geraniaceae, Oxalidaceae,
Sapindaceae and Vitaceae) of the Rosidae. The Rosales include
one CAM family, the Crassulaceae. Among the orders of the
Rosidae, the Rosales are most primitive and the other orders
are most likely derived from the Rosales (Cronquist, 1968).
The Celastraceae is considered one of the basic families
of the Celastrales. The Celastrales may be the origin for
the Euphorbiales (Cronquist, 1968). The Euphorbiaceae is a
large and highly diverse family. Cronquist (1968) feels that
the Rhamnales, Celastrates and Sapindales all may have
originated from a single complex in the Rosales. The
Geraniales appear to be derived from the Sapindales. Within
the Geraniales the Geraniaceae and the Oxalidaceae are
closely related.

Species with CAM occur in four families of the Asteridae
(Apocynaceae, Asclepiadaceae, Asteraceae and Labiatae). The
Apocynaceae and Asclepiadaceae which are closely related and
intergrade, are in the Gentianales. The Gentianales are the
most primitive order of the Asteridae. The other two CAM
orders, the Asterales and Lamiales, are thought to be
indirectly derived from the Gentianales. Cronquist (1968)
suggests the Gentianales gave rise to the Polemoniales which
gave rise to the Lamiales. He also suggests that the
Rubiales are derived from the Gentianales and the Asterales
are derived from the Rubiales. CAM has not yet been
unequivocally demonstrated in either the Polemoniales or the
Rubiales.

Within monocots, CAM occurs in two subclasses, the

Commelinidae and the Liliidae (classification on Cronquist,
1968). In the Commelinidae, the only CAM family thus far
described is the Bromeliaceae. This family is considered
to be a derived and terminal group within the subclass
(Cronquist, 1968), and many of the species are terrestrial
xerophytes or epiphytes.

The subclass Liliidae contains three CAM families, the
Agavaceae, Liliaceae and Orchidaceae. The phylogenetic
relationships of the genera and families are not fully
resolved at present. The Orchidaceae is clearly a highly
derived family and many species are tropical epiphytes.

POLYPHYLETIC ORIGINS

CAM occurs in four divisions of vascular plants, the
Lycophyta, Pteridophyta, Gnetophyta and Anthophyta. In
each of these four CAM divisions, the taxa that have CAM
represent a derived condition. At the level of the divisions,
therefore, the origin of CAM is polyphyletic. In the
Anthophyta CAM appears to have been independently derived in
5 of the 6 subclasses of the Magnoliatae. There are 17 orders
of vascular plants in which CAM is present. There are no data
available that suggest any of these orders do not also contain
C_3 species. However, some very large families, *e.g.*,
Cactaceae, Crassulaceae may be made up of species which
nearly all possess some type of CAM. Regardless of the
disputed relationships among many families it seems clear
that the occurrence of CAM in such distant families as the
Isoetaceae, Polypodiaceae, Welwitschiaceae, Agavaceae,
Asteraceae, Crassulaceae, Piperaceae and Cactaceae demands
the interpretation of a polyphyletic origin.

Within vascular plants C_4 photosynthesis occurs in only

one division, the Anthophyta. C_4 photosynthesis appears to have been independently derived in four subclasses of dicots and 1 subclass of monocots. At present 10 orders of vascular plants, all angiosperms, contain C_4 species. All 10 orders also contain C_3 species. The occurrence of C_4 photosynthesis in such distant families as the Polygonaceae, Cyperaceae, Asteraceae, Amaranthaceae, Euphoriaceae and Zygophyllaceae strongly suggests a polyphyletic origin.

Four orders of flowering plants (Caryophyllales, Euphorbiales, Geraniales, Asterales) are known to include both CAM and C_4 taxa. At the level of the family both CAM and C_4 are polyphyletic. In some families and in some genera it is possible that CAM or C_4 may also have evolved more than once (*e.g.*, Smith, 1976). However in most such families and genera the understanding of the phylogeny remains incomplete at present. For example, in the Gramineae as many as 6 independent origins of C_4 have been postulated (Smith, 1976). However, an equally good possibility remains that the ultimate interpretation of the phylogeny of the Gramineae could indicate only a single origin of C_4. Thus we are in need of additional information concerning the phylogeny of C_4 and CAM taxa within individual families and genera in order to better understand the evolution of these metabolic pathways.

MOSSES AND LICHENS

At present, only a limited number of moss and lichen species have been surveyed for $\delta^{13}C$ composition. The mosses studied have included species native to both arid and non-arid habitats. The mean $\delta^{13}C$ of these mosses was $-26°/_{oo}$ and all species had values that were more negative than $-21°/_{oo}$ (Rundel et al., 1979b; Teeri, in press), suggesting mosses utilize RuBP carboxylase in the uptake of atmospheric CO_2.

A comparison of the $\delta^{13}C$ values of six species of lichens
in Michigan yielded a mean $\delta^{13}C$ value of $-23.7°/_{oo}$ ($\sigma = 0.98$).
This value was significantly more positive than the mean $\delta^{13}C$
value of $-26.1°/_{oo}$ ($\sigma = 1.08$) of seven moss species growing
nearby (Teeri, in press). It remains to be demonstrated
whether the more positive $\delta^{13}C$ value of the lichens indicates
some contribution of PEP carboxylase activity in photosynthesis,
or whether it is caused by some other physical or metabolic
fractionation. Regardless of the cause, it still appears that
most of the carbon in the lichen biomass is a result of
atmospheric CO_2 uptake by RuBP carboxylase.

CARBON ISOTOPE RATIOS IN THE CRASSULACEAE

A rapidly increasing body of evidence suggests that there
is a previously-unsuspected amount of genetically determined
variability in the photosynthetic carboxylation processes
among the species in certain plant families. One such example
is the stonecrop family, the Crassulaceae, in which both the
C_3 and CAM pathways occur. In this family the $\delta^{13}C$ values
of plants of 108 species growing undisturbed in their natural
habitats show a continuum (fig. 3), from species that are
isotopically similar to C_3 plants with mean $\delta^{13}C$ values of
ca $-27°/_{oo}$, to species that are isotopically similar to dark-
fixing CAM plants with biomass mean $\delta^{13}C$ values of ca $-13°/_{oo}$.
Between these two extremes, there are species with mean $\delta^{13}C$
values ranging over the entire scale. The distribution of
$\delta^{13}C$ values is bimodal with most species clustering either
near $-13°/_{oo}$ or $-27°/_{oo}$. When the 108 species are considered
by genera (fig. 4), it is apparent that each of the nine
genera exhibits a range of $\delta^{13}C$ values occupying only part of
the possible range. In two cases, *Diamorpha* and the North

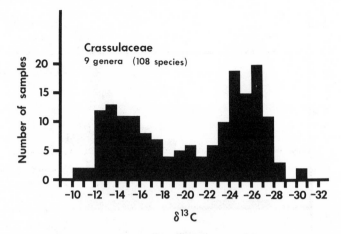

FIGURE 3.   *Frequency distribution of* $\delta^{13}C$ *values for mature leaf tissue of 108 species of the Crassulaceae. All plants were collected in their natural habitats and are representives of the following genera:* Crassula, Cremnophila, Diamorpha, Dudleya, Echeveria, Lenophyllum, Sedum, Sempervivum, *and* Villadia. *For several of the species samples of more than one population are included in the histogram. Sources of data are:* Crassula *(Mooney et al., 1977),* Cremnophila *(Teeri, in preparation),* Diamorpha *(Teeri, in preparation),* Dudleya *(Mooney et al., 1974; Troughton et al., 1977; Teeri, in preparation),* Echeveria *(Rundel et al., 1979a),* Lenophyllum *(Teeri, in preparation),* Sedum *(Teeri, in preparation),* Sempervivum *(Osmond et al., 1975)* Villadia *(Teeri, in preparation).*

American species of *Sedum,* the values are all more negative than $-20°/_{oo}$, suggesting that the sample plants primarily utilize the C_3 pathway in the uptake of atmospheric CO_2. At the other extreme, *Cremnophila, Echeveria* and *Lenophyllum* exhibit values all of which were more positive than $-16°/_{oo}$, suggesting that these plants mainly use CAM in taking up CO_2. Four genera, *Villadia, Dudleya, Sempervivum and Crassula,* have values falling between -11 or $-24°/_{oo}$. In these genera it appears that at least some species are utilizing both C_3

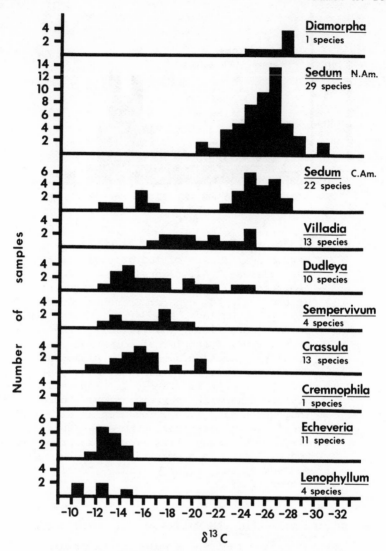

FIGURE 4. Frequency distribution of $\delta^{13}C$ values of mature leaf-tissue of species of nine genera of the Crassulaceae. The data and sources are the same as in Fig. 3. The genera are ranked in order of increasing mean $\delta^{13}C$ value.

and CAM pathways in the fixation of atmospheric CO_2.

Within species variability in the carbon isotope ratios among different populations is small (fig. 5). This limited

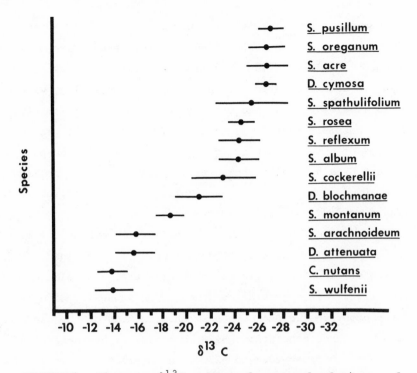

FIGURE 5. The mean $\delta^{13}C$ values of mature leaf tissue of
15 species in the Crassulaceae collected in
their natural habitats. The horizontal line
indicates one standard deviation above and
below the mean based on sample sizes of 3 to
12 populations per species. The genera are:
Sedum (S. pusillum, S. oreganum, S. acre,
S. spathulifolium, S. rosea, S. reflexum,
S. album, S. cockerellii); Diamorpha (D. cymosa);
Dudleya (D. blochmanae, D. attenuata); Sempervivum (S. montanum, S. arachnoideum, S. sulfenii);
Cremnophila (C. nutans). The sources of data are:
Kluge, 1977; Mooney et al., 1974; Osmond et al.,
1975; Troughton et al., 1977; and Teeri, in
preparation.

variability is illustrated by the populations of the 15 tested

species which had a mean σ of 1.7 and a mean range of $4°/_{oo}$.

One species, *Sedum spathulifolium,* had a standard deviation

of 3.0 for the five populations, much larger than that for

any of the other 14 species in figure 5. This is a complex,

wide-ranging species with four named subspecies (Clausen, 1975), two of which were included in this study. Perhaps the greater standard deviation reflects a greater degree of genetic differentiation of photosynthetic physiology within this species than in the other 14 species. Alternatively, it is possible that *S. spathulifolium* possesses a greater degree of phenotypic plasticity in its carboxylation pathways. Recent data have also shown that populations of *Sedum wrightii* from the southwestern U.S. and Mexico have a relatively broad range of $\delta^{13}C$ values of -20 to -26°/$_{\circ\circ}$ (Teeri, in prep.). A single sample from a plant of *S. wrightii* grown by D.R.T. Clausen in a greenhouse at Cornell University yielded a $\delta^{13}C$ value of -16.8°/$_{\circ\circ}$. Two other species that are related to *S. wrightii* also provide evidence of an unusually wide range of photosynthetic phenotypes in this section of the genus *Sedum*. Three populations of *S. cockerellii* had a range of $\delta^{13}C$ values from -20.4 to -25.4, and a greenhouse-grown plant of *Sedum caducum* was -17°/$_{\circ\circ}$. The genetic and environmental basis of this unusually broad range of $\delta^{13}C$ values for some species of *Sedum* clearly deserves further investigation.

In none of the 15 species in figure 5 is there isotopic evidence of a complete flexibility among populations in the production of biomass between daytime and nighttime atmospheric CO_2 fixation. Nor have the available isotopic values of other species in the Crassulaceae (figure 5) and *Welwitschia mirabilis* (Gnetaceae)(Schulze et al., 1976) and *Opuntia inermis* (Cactaceae)(Osmond et al., 1979) yielded any observations of complete flexibility under field conditions. Only in *Mesembryanthemum crystallinum* (Aizoaceae) has a relatively high degree of flexibility been demonstrated in the field (Winter et al., 1978). Thus, full flexibility between dark and light CO_2 uptake under field conditions appears to be an

Table 4. $\delta^{13}C$ *of leaf tissue of eleven species in their native habitats and grown together in a greenhouse*

Species	$\delta^{13}C$	
	Field	Greenhouse
Graptopetalum amethystinum	$-10.8°/_{oo}$	$-15.4°/_{oo}$
Echeveria agavoides	-12.9	-14.4
Sedum stahlii	-13.1	-14.7
Echeveria affinis	-13.3	-13.9
Echeveria pulidonis	-13.8	-13.5
Cremnophila nutans	-13.8	-12.9
Echeveria sanchez-mejorandae	-14.8	-15.3
Sedum dendroideum	-15.7	-13.7
Sedum spathulifolium		
ssp. pruinosum	-22.0	-25.1
Sedum greggii	-25.0	-25.1
Sedum ternatum	-26.4	-27.0

Sources of data: Rundel et al., 1979a, and Teeri, in preparation.

uncommon occurrence, at least as it is reflected in the production of photosynthetic tissue. These observations do not preclude the possibility that CO_2 fixation in the leaves of plants capable of CAM may at times occur solely by CAM pathways, even though those leaves were formed earlier from photosynthesis derived from C_3 photosynthesis. In such a case the $\delta^{13}C$ value of the structural carbon in the leaf tissue would be typical of C_3 photosynthesis even though only dark CO_2 uptake is occurring.

The differences in $\delta^{13}C$ values among the various species and genera considered here appear to be genetically determined. There are several types of evidence to support this. More closely related taxa have more similar $\delta^{13}C$ values, for example the different populations of a single species all have similar δ^{13} values (fig. 5). Interspecific differences in $\delta^{13}C$ values of field grown plants persist when those species are grown in the same greenhouse environment (table 4). An interesting

Table 5. Comparison of the Biomass $\delta^{13}C$ Values and Nocturnal
Accumulation of Titratable Acidity of Two Crassulacean Species
and Their F_1 Hybrid

	$\delta^{13}C$		Nocturnal Increase in Acidity
	Field	Greenhouse	
Sedum greggii	$-25°/_{oo}$	$-25°/_{oo}$	2-fold
Cremnophila linguifolia	-14	-13	4-fold
S. greggii x C. linguifolia		-19	4-fold

hybrid (table 5) further supports the suggestion of genetically determined differences in carboxylation among genera
of the Crassulaceae. *Sedum greggii* exhibits biomass $\delta^{13}C$
values indicative of daytime (C_3) photosynthesis both in the
field and greenhouse. This species accumulates acid at night,
and is a CAM plant according to the definition stated earlier,
even though CAM metabolism does not appear to be used in the
direct uptake of atmospheric CO_2. In contrast, *Cremnophila
linguifolia* has $\delta^{13}C$ values indicative of dark uptake of
atmospheric CO_2 both in the field and greenhouse. The F_1
hybrid of these two species, when grown in the greenhouse with
the two parents, has an intermediate $\delta^{13}C$ value of $19°/_{oo}$
indicating both carboxylation of atmospheric CO_2 is occurring
both in the day and night of this hybrid.

The evolutionary relationships among genera of the
Crassulaceae are incompletely understood. However, much
morphological and cytogenetic information is available for
many North and Central American genera. The genus *Sedum*
is divided here into North American species and Central
American species because extensive taxonomic studies have
suggested these are evolutionarily distinct groups (Clausen,
1959, 1975). A proposed phylogeny based on morphological

and cytogenetic criteria is presented in Fig. 6. In this
phylogeny, genera that are considered primitive have the
relatively negative $\delta^{13}C$ values of C_3 plants, and it is only
in the evolutionarily advanced genera that more positive $\delta^{13}C$
values occur. For example, the genus *Sedum* is primitive
and, in North America the most primitive subgenus of *Sedum*
is *Telephium* (Clausen, 1975) which is represented by a single
species *S. telephioides*. Three sampled populations of *S.
telephioides* yield a mean $\delta^{13}C$ value of $-25.8°/_{\circ\circ}$ (σ 1.5).
Both *Diamorpha* and *Parvisedum,* genera comprised of small
annual herbs, appear to be evolutionarily derived from *Sedum*
(Clausen, 1975). The photosynthetic physiology of these
genera has been little studied, but their very negative $\delta^{13}C$
values indicate that the biomass is derived from a daytime
carboxylation of atmospheric CO_2. Plants of both genera grow
in soils that can become very dry in summer, and their annual
habit appears to be the primary mechanism of adaptation to
these microenvironments. The North American species of
Parvisedum, Diamorpha and *Sedum* all have relatively thin leaves,
a character that appears to be strongly correlated with
relatively negative $\delta^{13}C$ values in the Crassulaceae.

A number of genera in the Crassulaceae appear to be
derived from the species of *Sedum* found in the Trans-Mexican
volcanic belt (Clausen, 1959, 1975). The large σ of 4.7 for
this group of *Sedum* reflects the fact that the full range of
possible $\delta^{13}C$ values has been measured among these species
(fig. 4). The morphological and cytogenetic evidence suggest
both *Cremnophila* and *Echeveria* are derived genera, possibly
from *Sedum* (Uhl, 1976). The 13 sampled species of these two
derived genera all yielded $\delta^{13}C$ values near $-13°/_{\circ\circ}$ (fig. 4).
The evolutionarily primitive *Sedum* is the most C_3-like in
$\delta^{13}C$ value and leaf thickness. The evolutionary advanced

Table 6. $\delta^{13}C$ *Values of Species of* Cremnophila, Diamorpha, Parvisedum *and* Sedum *in Their Native Habitats*

	Annuals and Biennials	Perennials
Number of species tested	9	34
Total number populations	20	55
Mean $\delta^{13}C$ (°/₀₀)	-26.5	-24.7
σ	1.4	2.7
Range	-24 to -30	-12 to -30

Phylogenies based on morphological and cytogenetic criteria

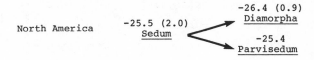

North America   -25.5 (2.0)
                 Sedum

-26.4 (0.9)
Diamorpha

-25.4
Parvisedum

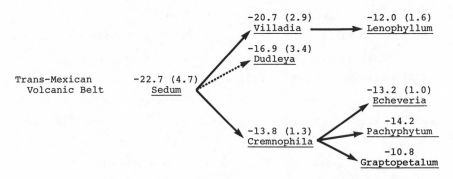

FIGURE 6. The mean $\delta^{13}C$ value of mature leaf tissue of plants of ten genera of the Crassulaceae. All plants were collected in their natural habitats. The arrows indicate increasing specialization of genera based on morphological and cyto- genetic criteria (see text). The standard deviation of the $\delta^{13}C$ value is in parentheses. In both Pachyphytum *and* Graptopetalum *there is only a single observation.*

genera *Echeveria* and *Cremnophila* are least similar to C_3
plants, with species of both genera having thick leaves and
apparently obligate dark fixation of atmospheric CO_2.
Increasing specialization of photosynthetic physiology toward
a predominance of dark-fixation in these taxa appears to
parallel increasing specialization as determined from cytogenetic and morphological criteria.

There appears to be a correlation between life cycle
length and the $\delta^{13}C$ value of studied species in the Crassulaceae (table 6). A survey of 34 randomly selected North
and Central American perennial species resulted in $\delta^{13}C$
values ranging from -12 to $-30°/_{oo}$. In contrast 9 species
that are either annual or biennial had values ranging from
-24 to $-30°/_{oo}$. Intuitively one would anticipate the shorter
life cycles of annuals and biennials to be associated with
relatively high growth rates and relatively high photosynthetic rates of C_3 plants. All of the studied annuals and
biennials had thin leaves, as is the case with perennial
species having $\delta^{13}C$ values in the range of -24 to $-30°/_{oo}$.
These observations are somewhat different from those reported
for the halophytic annual *Mesembryanthemum crystallinum*
(Aizoaceae)(Winter *et al.*, 1978). In this species, leaves
produced during the early, moist part of the growing season
had $\delta^{13}C$ values near $-26°/_{oo}$, exhibited no daily fluctuation
in organic acid content and had relatively high growth rates.
Leaves produced in the later, drier part of the growing
season had $\delta^{13}C$ values near $-16°/_{oo}$, exhibited substantial
diurnal fluctuation in organic acid content and had relatively
low growth rates. In these plants it appears that the decrease
in water availability is accompanied by the production of
leaves that exhibit an increasing proportion of dark uptake
of atmospheric CO_2 and presumably an increased water use

efficiency. Again it is intuitively evident that even for an annual species, the much lower growth rates associated with dark CO_2 uptake are potentially more advantageous than no growth at all at the dry end of the growing season. Further research is needed to determine to what extent the ratio of dark to light CO_2 uptake of a single fully expanded leaf can be altered by the onset of water stress. Additional information is needed as to the extent to which annual and biennial species in the Crassulaceae undergo seasonal shifts in dark uptake of atmospheric CO_2 and to what extent the recycling of respiratory CO_2 is altered over the growing season.

We are just beginning to understand the evolutionary and physiological variability associated with taxa capable of CAM. A major unanswered question concerns the extent to which differences among CAM taxa in $\delta^{13}C$ values are determined by environmental as opposed to genetically determined factors. The relationship between $\delta^{13}C$ values and the actual patterns of atmospheric CO_2 exchange throughout the life of the plant have only been investigated in a few species. Several studies have shown that some CAM taxa are extremely sensitive to the environment, such that the pathway used for CO_2 fixation is strongly influenced by environmental variables such as daylength, temperature and soil moisture availability (Kluge and Ting, 1978). The data presented here suggest that for a number of other species in the Crassulaceae environment has a relatively small effect on the type of carboxylation reactions used. In these species the available evidence suggests that differences among species in $\delta^{13}C$ values are genetically determined to a large extent. Additional information will be required to elucidate precisely how many genetically-determined ways plants incorporate CAM in their patterns of CO_2 metabolism. Such information will help to clarify the

relative importance of CAM as a means of producing new plant
biomass as compared to a means of recycling respiratory CO_2
during periods of unfavorable environmental conditions.

SPECULATION ON EVOLUTION

The biochemical and structural complexities of both C_4
and CAM photosynthesis present the problem of understanding
how such a complex set of properties could have evolved many
times. I will consider the case of CAM. The biochemical and
structural characteristics of CAM occur in distantly related
groups of vascular plants. Such a widespread occurrence can
by itself be interpreted as indicative of a primitive state.
If we refer to CAM according to the definition previously
stated, then there are both advanced vascular plants (*e.g.,*
angiosperm CAM plants) and primitive vascular plants (*e.g.,*
Isoetes, Welwitschia) that possess the biochemical and
structural characteristics required for the function of CAM
in the uptake of atmospheric CO_2. If the biochemical and
structural characteristics of CAM plants are indeed primitive
conditions among vascular plants, then it may well be that
the genes governing the expression and co-ordination of these
characteristics are present in all vascular plants, but not
necessarily expressed as part of the photosynthetic system.
The pathway may not function in the uptake of atmospheric CO_2,
or in the recycling of respiratory CO_2 in most vascular plants,
and thus they are classified as C_3 plants. If the pathway is
present in all vascular plants, then the evolution of CAM as
a means of photosynthetic uptake of atmospheric CO_2, or
recycling of respiratory CO_2, would then involve the turning
on or amplification of expression of genes that already exist
in the photosynthetic cells of C_3 plants. Such an evolutionary

process would require a much simpler set of genetic events
for the repeated appearance of CAM photosynthesis, than would
the repeated evolution, *de novo*, of all biochemical and
structural components of CAM in each of the many lineages in
which CAM has appeared.

A series of recent investigations (*e.g.*, Outlaw, 1981;
Schnabl, 1980; Thorpe *et al.*, 1978; Willmer, 1980) provides
several kinds of evidence for the similarity between CAM, as
defined above, and the organic acid metabolism of guard cells
of C_3 plants. Guard cells thus far studied fix CO_2 either in
the light or in the dark. Major products of the CO_2 fixation
in guard cells are C_4 organic acids, particularly malic acid.
The activity of PEP carboxylase is high in guard cells. There
is an inverse correlation between guard cell starch concentration and malic acid concentration, suggesting that
degradation of starch provides a precursor in the formation
of malic acid (Outlaw, 1980, 1981; Schnabl, 1980). Guard
cells undergo a several-fold increase in malic acid content
during stomatal opening (Outlaw, 1981). The decarboxylation
of malic acid in some, but not all CAM plants appears to
occur via NADP malic enzyme (Osmond, 1978) and high levels of
activity of this enzyme have been detected in *Vicia* guard
cells (Outlaw, 1981). The first part of the definition of a
CAM cell requires that a photosynthetic cell is capable in the
dark of fixing CO_2 via PEP carboxylase into malic acid which
accumulates in the cell. Guard cells thus far studied have
this capability and the magnitudes of the changes in malic
acid content of guard cells appear to be comparable to that
of many CAM cells over a daily cycle. The second part of the
definition of a CAM cell requires that in the following light
period the malic acid can be decarboxylated and the CO_2 enters
the PCR cycle in the same cell. It is clear that guard cells

undergo decrease in malic acid content during stomatal closure.
However, unlike CAM cells, recent evidence suggests that the
PCR cycle is not functional in guard cell chloroplasts (Thorpe
et al., 1978; Schnabl, 1980; Outlaw, 1981). Although guard
cells do contain chloroplasts, Outlaw (1981) has found that
several enzymes unique to the PCR cycle, including RuBP
carboxylase, were not active in studied guard cells of *Vicia
faba*. Outlaw (1981) suggested that there may be an incompatibility between starch degradation, which occurs in the
light of guard cells and the operation of an active PCR cycle
in the same cell. In a CAM cell the PCR cycle is active in
the light whereas starch is degraded in the dark. Thus any
incompatibility between these two processes in a CAM plant
may be minimized, or eliminated, by the temporal compart-
mentation of the two processes.

The chloroplasts of CAM plants exhibit a range of granal
structures from those termed "granal" to those termed "agranal"
(Kluge and Ting, 1978). No structures unique to CAM plant
chloroplasts have been reported. However, there appear to be
quantitative differences in the chloroplast structure of CAM
plants in comparison to many C_3 plants. The most frequently
reported difference is that the granal stacks of many CAM
plants are reduced in size and contain smaller number of
thylakoids per granum than in many C_3 plants. Guard cells
of C_3 plants also have been reported to have reduced numbers
of thylakoids per granal stack. In the CAM plant *Cremnophila
linguifolia* there were an average of 3.8 thylakoids per granum
(Teeri and Overton, in press), similar to the chloroplasts of
guard cells of two C_3 species which both had an average of
3.9 thylakoids per granal stack (Allaway and Setterfield,
1972).

At present, there are too few data to permit a rigorous

comparison of CAM cells and guards cells. Many of the cited
similarities could be correlative but not necessarily causal
relations.

The hypothesis suggested by the preceding is that the
evolution of CAM involves the derepression of genes that are
present, but not normally expressed, in the photosynthetic
cells of C_3 plants. One possible reason for the ubiquitous
occurrence of such genes would be that they may normally
function in guard cell metabolism of all C_3 species. If this
hypothesis were true, then the evolution of CAM would involve
genetic change primarily concerned with the regulation of the
proportion of C_3 and C_4 carboxylation reactions in photosynthetic cells. It is also possible that pathways of carbon
metabolism similar to CAM exist at other points of metabolism.
Yet the similarities between guard cells and CAM cells are
provocative and clearly the subject deserves further study.

Present evidence from the Crassulaceae strongly suggests
a genetic basis for the quantitative variability that exists
in $\delta^{13}C$ values among the various genera and species. The
metabolic machinery necessary for the operation of CAM may
represent a widespread and primitive character among vascular
plants. The use of this metabolic machinery in the uptake of
atmospheric CO_2 appears to represent a derived condition in
the various groups of vascular plants where it occurs.
Additional information is required to determine if a similar
origin is to be suggested for the genes governing C_4 photosynthesis.

Finally, I wish to point out the very great value of
herbarium material in this kind of research. The availability
of large collections of properly identified samples has been
the major source of data for several recent advances in
understanding of the ecological and genetic variability that

exists both in CAM and C_4 families. As new physical and chemical techniques are developed, I am confident there will be an increasing contribution of the resources of these collections to investigations in evolutionary plant biology.

ACKNOWLEDGMENTS

I thank Drs. R.T. Clausen and C. Uhl for much assistance in my studies of the Crassulaceae, and Dr. A. Naylor who originally stimulated my interest in C_4 and CAM. Dr. W. Burger generously made available the resources of the herbarium of the Field Museum of Natural History. Excellent technical assistance has been provided by Ms. E. McCarthy, Mr. W. Schroeder, Mr. E. van Santen, Ms. S. Yamins and Ms. T. Teeri. Drs. M. Darling, J. Keeley, S. Szarek and W. Outlaw, Jr., provided helpful comments on the manuscript. I thank my many colleagues who have discussed various aspects of this research with me over the past two years. This research was supported in part by DeKalb AgResearch, Inc.; NSF DEB-8021270, NSF DEB-8021312 (to the Duke University Phytotron) and NIH AM-18741.

LITERATURE CITED

ALLAWAY, W.G., and G. SETTERFIELD. 1972. Ultrastructural observations on guard cells of *Vicia faba* and *Allium porrum*. Canadian J. Bot. 50:1405-1413.
BENDER, M.M. 1968. Mass spectrometric studies of carbon 13 variation in corn and other grasses. Radiocarbon 10:468-472.
BENDER, M.M. 1971. Variations in $^{13}C/^{12}C$ ratios of plants in relation to the pathway of photosynthetic carbon fixation. Phytochemistry 10:1239-1244.
CLAUSEN, R.T. 1959. *Sedum* of the Trans-Mexican volcanic belt. Ithaca: Cornell University Press. Pp. 1-380.

128   *James A. Teeri*

CLAUSEN, R.T. 1975. *Sedum* of North America north of the
Mexican Plateau. Ithaca: Cornell University Press.
Pp. 1-742.
CLIFFORD, H.T., and W.T. WILLIAMS. 1980. Interrelationships
amongst the Liliatae: A graph theory approach. Aust. J.
Bot. 28:261-268.
CRAIG, H. 1953. The geochemistry of the stable carbon
isotopes. Geochimica et Cosmochimica Acta 3:53-92.
CRONQUIST, A. 1968. The evolution and classification of
flowering plants. Boston: Houghton Mifflin Co. Pp. 1-396.
KEELEY, J.E. 1981. Isoetes howellii: A submerged aquatic
CAM plant? American Journal of Botany 68:420-424.
KLUGE, M. 1976. Models of CAM regulation. In: R.H. Burris
and C.C. Black, eds., CO_2 metabolism and plant productivity.
University Park Press. Pp. 205-216.
KLUGE, M. 1977. Is *Sedum acre* L. a CAM plant? Oecologia
29:77-83.
KLUGE, M., and I.P. TING. 1978. Crassulacean Acid Metabolism.
Berlin: Springer Verlag. Pp. 1-209.
MOONEY, H., J.H. TROUGHTON and J.A. BERRY. 1974. Arid climates
and photosynthetic systems. Carnegie Inst. Yearbook
73:793-805.
MOONEY, H.A., J.H. TROUGHTON, and J.A. BERRY. 1977. Carbon
isotope ratio measurements of succulent plants in
southern Africa. Oecologia 30:295-305.
NAMBUDIRI, E.M.V., W.D. TIDWELL, B.N. SMITH, and N.P. HEBBERT.
1978. A C_4 plant from the Pliocene. Nature 276:816-817.
OSMOND, C.B. 1978. Crassulacean acid metabolism: a curiosity
in context. Ann. Rev. Plant Physiol. 29:379-414.
OSMOND, C.B., O. BJÖRKMAN, and D.J. ANDERSON. 1980.
Physiological processes in plant ecology. Berlin: Springer-
Verlag. Pp. 1-468.
OSMOND, C.B., D.L. NOTT, and P.M. FIRTH. 1979. Carbon
assimilation pattern and growth of the introduced CAM
plant *Opuntia inermis* in eastern Australia. Oecologia
40:331-350.
OSMOND, C.B., H. ZIEGLER, W. STICHLER, and P. TRIMBORN. 1975.
Carbon isotope discrimination in alpine succulent plants
supposed to be capable of Crassulacean acid metabolism
(CAM). Oecologia 18:209-217.
OUTLAW, W.H., Jr. 1980. A descriptive evaluation of
quantitative histochemical methods based on pyridine
nucleotides. Ann. Rev. Plant. Physiol. 31:299-311.
OUTLAW, W.H., Jr. 1981. Unique aspects of carbon metabolism
in guard cells of *Vicia faba* L. In: Integrated View of
Guard Cells, ed., C.A. Rogers, pp. 4-18. Proceedings
of a Symposium of the Southern Section of the American
Society of Plant Physiologists, Greenville, Mississippi.

RAO, I.M., P.M. SWAMY, and V.S.R. DAS. 1979. Some
characteristics of Crassulacean acid metabolism in
five nonsucculent scrub species under natural semiarid
conditions. Z. Pflanzenphysiol. 94:201-210.

RUNDEL, P.W., J.A. RUNDEL, H. ZIEGLER, and W. STICHLER. 1979a.
Carbon isotope ratios of central Mexican Crassulaceae in
natural and greenhouse environments. Oecologia 38:45-50.

RUNDEL, P.W., W. STICHLER, R.H. ZANDER, and H. ZIEGLER. 1979b.
Carbon and hydrogen isotope ratios of bryophytes from
arid and humid regions. Oecologia 44:91-94.

SCHNABL, H. 1980. CO_2 and malate metabolism in starch-
containing and starch-lacking guard-cell protoplasts.
Planta 149:52-58.

SCHULZE, E.-D., H. ZIEGLER, and W. STICHLER. 1976. Environmental control of Crassulacean acid metabolism in
Welwitschia mirabilis Hook. Fil. in its range of natural
distribution in the Namib desert. Oecologia 24:323-334.

SMITH, B.N. 1976. Evolution of C_4 photosynthesis in response
to changes in carbon and oxygen concentrations in the
atmosphere through time. BioSystems 8:24-32.

STOWE, L.G., and J.A. TEERI. 1978. The geographic distribution
of C_4 species of the Dicotyledonae in relation to climate.
American Naturalist 112:609-623.

SZAREK, S.R. 1979. The occurrence of Crassulacean acid
metabolism: A supplementary list during 1976 to 1979.
Photosynthetica 13:467-473.

TAKHTAJAN, A. 1969. Flowering plants: Origin and dispersal,
transl. C. Jeffrey. Washington, D.C.: Smithsonian
Institution Press. Pp. 1-310.

TAYLOR, T.N. 1981. Paleobotany: An introduction to fossil
plant biology. New York: McGraw-Hill Book Co. Pp. 1-589.

TEERI, J.A. 1979. The climatology of the C_4 photosynthetic
pathway. In: O.T. Solbrig *et al.*, eds., Topics on
plant biology. Columbia University Press. Pp. 356-374.

TEERI, J.A., and L.G. STOWE. 1976. Climatic patterns and
the distribution of C_4 grasses in North America. Oecologia
23:1-12.

TEERI, J.A., L.G. STOWE, and D.A. MURAWSKI. 1978. The
climatology of two succulent plant families: Cactaceae
and Crassulaceae. Canadian Journal of Botany 56:1750-1758.

TEERI, J.A., L.G. STOWE, and D.A. LIVINGSTONE. 1980. The
distribution of C_4 species of the Cyperaceae in relation
to climate. Oecologia 47:303-310.

TEERI, J.A. Stable carbon isotope analysis of mosses and
lichens growing in xeric and moist habitats. Bryologist
(in press).

TEERI, J.A. and J. OVERTON. Chloroplast ultrastructure in two
   Crassulacean species and an F_1 hybrid with differing
   biomass $\delta^{13}C$ values. Plant, Cell and Environment (in
   press).
THORPE, N., C.J. BRADY, and F.L. MILTHORPE. 1978. Stomatal
   metabolism: Primary carboxylation and enzyme activities.
   Aust. J. Plant. Physiol. 5:485-493.
TROUGHTON, J.H. 1979. $\delta^{13}C$ as an indicator of carboxylation
   reactions. In: M. Gibbs and E. Latzko, eds., Photosynthesis
   II: Photosynthetic Carbon Metabolism and Related Processes.
   Berlin: Springer-Verlag. Pp. 140.
TROUGHTON, J.H., H.A. MOONEY, J.A. BERRY, and D. VERITY. 1977.
   Variable carbon isotope ratios of *Dudleya* species in
   natural environments. Oecologia 30:307-311.
UHL, C.H. 1976. Chromosomes, hybrids and ploidy of *Sedum
   cremnophila* and *Echeveria linguifolia* (Crassulaceae). Am.
   Jour. Bot. 63:806-820.
WILLMER, C.M. 1980. Some characteristics of phosphoenol-
   pyruvate carboxylase activity from leaf epidermal tissue
   in relation to stomatal functioning. New Phytol. 84:593-602.
WINTER, K. 1981. C_4 plants of high biomass in arid regions of
   Asia -- occurrence of C_4 photosynthesis in Chenopodiaceae
   and Polygonaceae from the Middle East and USSR. Oecologia
   48:100-106.
WINTER, K., U. LÜTTGE, E. WINTER, and J.H. TROUGHTON. 1978.
   Seasonal shift and C_3 photosynthesis to Crassulacean acid
   metabolism in *Mesembryanthemum crystallinum* growing in its
   natural environment. Oecologia 34:225-237.
ZIEGLER, H., K.H. BATANOUNY, N. SANKHLA, O.P. VYAS, and W.
   STICHLER. 1981. The photosynthetic pathway types of some
   desert plants from India, Saudi Arabia, Egypt and Iraq.
   Oecologia 48:93-99.

CHEMICAL CONSTITUENTS OF NECTAR IN RELATION
TO POLLINATION MECHANISMS AND PHYLOGENY

Herbert G. Baker and *Irene Baker*

Department of Botany
University of California
Berkeley, California

Floral nectar is one of the rewards that angiosperms make
available to animal visitors (insects, birds or bats) to their
flowers that may be useful as pollen-vectors. Consequently,
its chemistry is important biologically. Nectars from over
1000 species of flowering plants have been analyzed in our
studies of various classes of nectar constituents. Sugars
and amino acids have been identified individually and their
quantities measured; other chemicals were treated on a
"presence or absence" basis. Sugar concentrations in nectars
are related to pollinator types although subject to environmental influence; sugar ratios (sucrose/glucose + fructose)
are less affected environmentally and show a superimposition
of pollinator adaptation on phylogenetic relationship. Nectars
of flowers adapted to hummingbirds, Lepidoptera, long-tongued
bees, and wasps, tend to have high sucrose/hexose ratios;
nectars of flowers adapted to passerine (perching) birds, bats
(at least Microchiroptera), flies, short-tongued bees and "bees
and butterflies" tend to have low ratios. In Penstemon
(Scrophulariaceae) and Erythrina (Fabaceae) there is clear
evidence of pollinator adaptation of sugar ratios within a
genus, but in the Asteraceae there is evidence of "phylogenetic
constraint".

Nectar amino acid concentrations are adaptive, with a
tendency for higher concentrations in taxa whose visitors have
no alternative source of protein-building material. However,
nectar of flowers luring carrion- and dung-flies to them are
exceptionally strong in amino acids. Amino acid complements
have taxonomic (and, therefore, phylogenetic) significance
(exemplified by species of Penstemon) but are less obviously

*selected by pollinator needs, showing strong phylogenetic
constraint. Inheritance of amino acid complements in hybrids
between closely related species is additive in the F₁
generation, but shows segregation in subsequent generations.*

*Attention is also given to the occurrence in nectar of
lipids, as well as to "organic acids" and proteins (probably
usually enzymes), all of which may be related to selection by
pollinators. The pH of nectar ranges from at least pH 3 to
pH 10, but the significance of this is not yet elucidated.
Among substances in nectar that may be deterrents to
inappropriate flower visitors we have tested for, and found,
non-protein amino acids, alkaloids and phenolics. They are
more common in tropical floras than in those of temperate
and alpine regions.*

*The importance of conducting future work on nectar chemistry
on an ecosystem basis, rather than the consideration of individual species* in vacuo, *is emphasized.*

Floral nectar is one of the rewards that flowers make
available to those visitors that may be useful to the plant
as pollen vectors. The other obvious reward is pollen (over
and above that which is required for the act of pollination).
Traditionally, it has been believed that bees, in particular,
utilize nectars as a source of energy-producing sugars for
their own use (and that of their brood), while pollen is
collected primarily for its protein-building potentiality
(Haydak, 1970).

However, with the appreciation that a much wider range
of insects visit flowers, and that other animals -- birds,
bats and rodents -- may be pollinatory flower-visitors, the
situation is clearly more complicated. This is aside from
the facts of wind-pollination for many plants (and water-
pollination for a few) and an emphasis on self-pollination
or apomixis in many others.

Consequently, in the last decade we, and a growing
number of other interested persons, have been looking at
nectar and pollen chemistry to see what may be found in the

way of pollinatory adaptations and to see if phylogenetic
relationships are reflected in the chemical constitutions
of the nectar and pollen. We should emphasize that we are
biologists -- nearly pure and rather simple -- and we do not
claim expertise in biochemistry, but we have the capability
of repeating simple tests and using more complex analytical
methods in wide-ranging surveys, as well as more intensive
studies of particular taxa.

For reasons of limited space-availability, we shall
concentrate on nectar rather than pollen. We believe that
the fact that nectar is simply a reward and, unlike pollen,
is not complicated by also being the agent of delivery of
the male gametes to the gynoecium, means that more obvious
selective effects in relation to pollinator type may be found
for nectar.

With some insects relying entirely upon nectar as a food-
source for the adult organism (most obviously Lepidoptera and
some Diptera), it might be expected that there would be more
in nectar than just sugar and water, and this has been found
to be the case. A wide range of chemicals occurs in nectars
of various species (table 1 -- for details and references,
see Baker, H.G. and I. Baker, 1975, p. 102), although sugars
are by far the largest components.

With tests for some of these substances in mind, freshly-
produced nectars from over 1000 species of flowering plants
have been tested or analyzed for one or more classes of
compounds. Most emphasis has been given to the sugars and
amino acids and their presence (which is universal) has
always been estimated quantitatively. Other chemicals have
been estimated quantitatively in a majority of cases (e.g.,
proteins) while the remainder (lipids, antioxidants, phenolics
and alkaloids), so far, have been tested for on a "presence

or absence" basis only. Our methods are published in
appropriate detail elsewhere (sugars in Baker, H.G., and
I. Baker, 1979, 1981; amino acids in Baker, I. and H.G. Baker,
1976, and with modifications in Baker, *et al.*, 1978; proteins,
lipids, antioxidants and alkaloids in Baker, H.G. and I. Baker,
1975; phenolics in Baker, H.G. 1977, 1978; Baker, I. and H.G.
Baker, 1981).

SUGARS

Although a wide variety of sugars has been identified in
nectars by various investigators (Baker, H.G. and I. Baker,
1981), they all agree that the disaccharide sucrose and the
hexose monosaccharides glucose and fructose are the commonest
and, when they occur, the most concentrated. Among the

Table 1. *Chemicals reported to be present in the nectars*
          of flowering plants (from Baker, H.G. & I. Baker,
          1975)

Sugars

Amino acids
Proteins
Lipids
Phenolic compounds
Alkaloids
Glycosides

Thiamin, Riboflavin, Nicotinic acid, Pantothenic acid,
     Pyridoxin, Biotin, Folic acid, Mesoinositol.

"Organic acids" - Ascorbic, Fumaric, Succinic, Malic,
     Oxalic, Citric. Tartaric, α-ketoglutaric, Gluconic,
     Glucuronic, etc.

Allantoin and Allantoic acid

Dextrins

Inorganic substances

---

remainder (table 2) only the disaccharide maltose and the
trisaccharide melezitose are at all frequently occurring in
fresh (or freshly dried) nectar. All our analyses and
estimations of the sugars are done by descending paper chromatography from nectar samples dried immediately after collection
and eluted when needed (Baker, H.G. and I. Baker, 1979a, 1981).

Sugar Concentrations

The general assumption of biologists who study the
energetics of flower-visitors that imbibe nectar is that a
satisfactory picture of sugar concentration can be developed
by relying on the refractive index of the nectar (which is
referred to in terms of "sucrose equivalents"). This assumes
that sucrose, glucose and fructose (which have roughly equal
effects on the refractive index on a weight basis) account
for almost the entire amount of sugar present -- a supposition

---

*Table 2.   Minor sugars reported from nectar
            (references in Baker, H.G. & I. Baker, 1981)*

Monosaccharides
    arabinose
    galactose
    mannose
Disaccharides
    gentiobiose
    lactose
    maltose
    melibiose
    trehalose
Trisaccharides
    melezitose
    raffinose
Tetrasaccharides
    stachyose

---

that is no more inaccurate than many of the other assumptions
upon which nectar-feeder energetics are based. However,
Inouye *et al.* (1980) have shown that lipids, ascorbic acid
and amino acids (that can be present in nectar -- Baker, H.G.
and I. Baker, 1975; Baker, H.G. 1977, 1978) may affect the
refractive index significantly (although not immodestly at
concentrations that most often occur naturally).

In general, the concentration of sugars in nectars that
are imbibed through narrow tubes by the flower-visitors (such
as Lepidoptera) are lower (and, consequently, less viscous)
than the nectars utilised by bees and by flies, which have
less narrow mouthparts (table 3).   Hummingbirds, a portion
of whose tongue fills by capillarity, and microchiropteran
bats that spend only a second or less at a flower, also select
for relatively dilute, low-viscosity nectar (Baker, H.G., 1975).
Many kinds of insects regurgitate liquid onto concentrated or
crystallized nectar and, by doing so, dilute it so that it
can be taken up.

This raises the point that nectar, once produced and
exposed to the air, will tend to evaporate, particularly in
bowl-shaped flowers exposed to sun and wind, and it has been
suggested by Dr. F.L. Carpenter (pers. comm.) that the
tubular shape of most hummingbird-flowers (and, it might be
added, of hawkmoth-flowers) is more important as an
evaporation-reducing device than as a guide to the beaks of
birds leading to the nectar accumulated at the base of the
tube (see, also, Corbet, 1978).   However, it must always be
remembered that the hiding away of nectar from short-tongued
nectar-thieves is an important function of a tubular corolla.

Preliminary data that we have accumulated from lowland
dry tropical forest and from cloud forest (both in Costa Rica),
and from the subalpine and alpine zones of the Colorado Rockies,

Table 3. Sugar concentrations of nectars (on a sucrose-equivalent, weight per total weight basis) of plants of various pollinator classes in lowland dry forest, cloud forest (both in Costa Rica) and the alpine and subalpine zones of the Rocky Mountains (in Colorado).

| | Lowland dry forest | | Cloud forest | | Alpine + subalpine | |
| | Costa Rica | | Costa Rica | | Colorado | |
	(N)	x̄(%)	(N)	x̄(%)	(N)	x̄(%)
Short-tongued bees + flies	(5)	46	(19)	21	(41)	32
Long-tongued bees	(15)	46	(12)	24	(27)	37
Bees + butterflies	–		(8)	22	(26)	27
Butterflies	(20)	29	(7)	14	(4)	24
Settling moths	(2)	41	(8)	18	(1)	18
Hawkmoths	(11)	24	(8)	15	–	
Bats	(5)	17	(4)	15	–	
Hummingbirds	(29)	21	(20)	17	(3)	41
	(87)		(86)		(102)	

(Baker, H.G., 1978 and Baker, H.G. and I. Baker, in prep),
make it clear that the absolute values for nectar sugar
concentrations will vary with climatic conditions (table 3).
Thus, in the trees and vines that flower in the dry season
in the lowland tropical forest, much higher concentrations
are reached than in the flowers living in the highly humid,
cool conditions of the cloud forest. Also the restricted
volumes of nectar in the Colorado alpine zone contain high
concentrations of nectar sugars. This was also noticed in
the arctic tundra by Hocking (1953). However, the relative
concentrations of the various pollinator classes tend to
remain in the same order.

Sugar Ratios

    In our major studies we have been more concerned with the
representation of individual sugars in nectar than with their
overall concentration. In particular, we believe that there
are strong correlations between the ratio of sucrose to
hexoses and pollinator types. Additionally, these ratios
show within-family resemblances that are phylogenetically
interesting. Our studies have shown that despite environmentally produced variations in amount and concentrations of
nectar, the sugar ratios remain remarkably constant. They
are also consistent between individuals, even in obligately
out-crossed species (Baker, H.G. and I. Baker, 1979, 1981;
Baker, H.G., 1978). Such consistency may be important in
giving the nectar a uniform "taste," recognizable by a
visitor to the flowers (Baker, H.G., 1978). The minor sugars
(table 2) may contribute to this "taste."
    In reporting our studies of sugar ratios (Baker, H.G.,
1978; Baker, H.G. and I. Baker, 1979a, 1981), we have adopted

a terminology which recognizes four classes of nectar. These
are: "sucrose-dominant" when the sucrose/glucose + fructose
ratio is greater than 0.999; "sucrose-rich" for nectars with
ratios between 0.5 and 0.999; "hexose-rich" for those with
ratios between 0.1 and 0.499; and "hexose-dominant" for those
with ratios less than 0.1. In some previous publications
(Baker, H.G., 1978; Baker, H.G. and I. Baker, 1979a; Baker,
I. and H.G. Baker, 1979), ratios less than 0.5 were called
"sucrose-poor," but we believe now that it is more appropriate
to refer to them as "hexose-rich" or "hexose-dominant."

Data on the occurrence of the "big three" sugars from
nectars of 765 species (growing in many parts of the temperate
and tropical world) are summarized qualitatively in table 4.
Most nectars contain sucrose, glucose and fructose. No
nectars were found that contain only fructose or sucrose +
fructose, but all other combinations occur.

Percival (1961), in a large survey (over 800 species) that
was only subjectively semi-quantitative, remarked that some
flowering plant families may be characterized by sucrose-rich
nectars while others are predominantly or entirely hexose-
rich. Our findings (Baker, H.G. and I. Baker, 1981)
corroborate this statement. Thus, the Brassicaceae (Cruciferae)

Table 4.   *Numbers of nectars with detectable sugar-
            combinations*

Sucrose	7
Glucose	2
Fructose	0
Sucrose + glucose	29
Sucrose + fructose	0
Glucose + fructose	78
Sucrose + glucose + fructose	<u>649</u>
	765

and Asteraceae (Compositae) are predominantly hexose-rich
while the Ranunculaceae and Lamiaceae (Labiatae) are predominantly sucrose-rich families. Other families, such as
the Scrophulariaceae, Bignoniaceae, and the Fabaceae
(Leguminosae) show a very wide range of sugar-ratios, even
within genera; these ranges of sugar ratios can be correlated with a range of pollination modes. Harborne (1977,
p. 52) reports some observations on the nectar sugars of
Rhododendron where he recognized five classes depending on
the proportions of sucrose, glucose and fructose along with
an oligosaccharide (possibly raffinose). This categorization
followed sectional groupings within the genus, so that its
taxonomic potentiality is apparent.

Despite suggestions in the past (Wykes, 1952a, 1952b) that
honey bees prefer a "balanced" nectar, with equal quantities
of the three main sugars, more recent studies (Percival, 1961;
Waller, 1972) suggest that this is not the case. Certainly,
our results (Baker, H.G. and I. Baker, 1981) show that honey
bees are provided with nectars of a wide range of sugar ratios
by flowers of different species.

Hummingbirds are the only other flower-visiting nectar-
consumers whose preferences have been tested. The works of
van Riper (1960), Hainsworth and Wolf (1976) and Stiles (1976)
show that these birds do have a preference -- for nectars
that contain a relatively high proportion of sucrose (see
review in Baker, H.G. and I. Baker, 1981). With at least
this evidence of a connection between sugar ratio and
pollinator type, we decided to analyze all our nectar data
on the basis of principal pollinator classes. The results
of the analyses are shown in table 5. The total for the
different classes adds up to more than 765 because some
species have more than one important pollinator.

Table 5. Numbers of species in each of four sugar-ratio categories arranged by principal pollinators. Data from Baker, H.G. and I. Baker (1981).

| | $\frac{S}{G+F}$ | | | | | | |
	<0.1	0.1 to 0.499	0.5 to 0.999	>0.999	N	G*	P**
OVERALL	195	231	149	190	765	-	-
Hummingbirds	0	18	45	77	140	119.52	<.001
New World Passerines	11	1	0	0	12	25.16	<.001
Sunbirds, etc.	24	9	2	0	35	28.07	<.001
Honeyeaters	18	4	0	0	22	36.87	<.001
Honeycreepers	5	1	0	0	6	10.57	<.02
Lorikeets, etc.	1	2	0	0	3	3.69	.30
Hawkmoths	2	8	19	32	61	41.16	<.001
Settling moths	3	14	11	15	43	70.07	<.001
Butterflies & skippers	5	17	24	29	75	24.23	<.001
Short-tongued bees + butterflies	23	21	3	0	47	38.07	<.001
Short-tongued bees	115	103	28	17	263	75.47	<.001
Long-tongued bees	13	75	49	66	203	42.40	<.001
New World bats	9	18	0	0	27	32.51	<.001
Old World bats	1	3	2	1	7	1.36	.90
Non-volant mammals	0	2	2	1	5	13.44	<.01
Wasps	2	7	4	5	18	1.24	.75
Beetles	1	3	2	3	9	1.22	.75
Flies	29	27	7	9	72	14.82	<.001

*G = G statistic (see Sokal and Rohlf, 1969, chapter 16)
**P = probability of difference from OVERALL

1) Bird-pollinated flowers -- Hummingbird-flowers have
high sucrose/glucose + fructose ratios, while flowers
pollinated primarily by passerine (perching) birds have very
low ratios. The passerine birds included in this category of
flower nectar-takers include sunbirds (Nectariniidae) and
white-eyes (Zosteropidae) in Africa and Asia, honeyeaters
(Meliphagidae) mostly in Australia, honey creepers (Drepanididae
in the Hawaiian Islands, as well as a variety of perching
birds in the New World tropics and sub-tropics (Icteridae,
Thraupidae, Parulidae, etc.).

The striking contrast between the nectars of hummingbirds
(Trochilidae) and those of passerine birds can be seen in
table 5 on a general basis, but it can also be illustrated
by the species of a particular genus, *Erythrina* (Fabaceae).
This genus of trees is found in the tropics and subtropics
of both the New and the Old Worlds. Old World and New World
passerine bird-pollinated species of this genus show clear
hexose-dominance while the hummingbird-pollinated species
(all from the New World, of course) are never less than
"sucrose-rich" (table 6)(Baker, H.G. and I. Baker, 1981;
Baker, I. and H.G. Baker, 1981).

The physiological basis of the actual preference by
hummingbirds for sucrose-rich nectar, and the no-less likely
preference of passerine birds for nectars rich in hexose
sugars, is unclear at present. Sucrose, glucose and fructose,
weight for weight, differ only slightly in caloric value,
suggesting that energetics are not involved but that a matter
of "taste" is involved. For passerine birds there may be
a clue in the fact that many of them also feed at juicy or
fleshy fruits, as well as being nectarivorous. In our
experience (Baker, H.G. and I. Baker, unpubl.), these fruit
juices are almost always stronger in hexose sugars than in
sucrose. Consequently, the passerine birds may have developed

"taste search-image" which can be satisfied by "hexose-
ominated" or "hexose-rich" nectars. Hummingbirds, which
o not usually feed on fruits, would not make this association
nd, in any case, hummingbird-flowers are believed to be often
erived from large bee-pollinated ancestors (Grant and Grant,
968) and, as we shall point out later, the nectars of large
ee flowers are usually sucrose-rich.

    2) Bat-pollinated flowers -- Flower-visiting bats in the
eotropics belong to the suborder Microchiroptera (fig. 1), and
he flowers they visit have nectar that is hexose-dominated or
exose-rich (table 5)(Baker, H.G., 1978; Baker, H.G. and I.
aker, 1981; Scogin, 1980), but the flowers visited by bats in
he Paleotropics (suborder Megachiroptera) (fig. 2) appear on the
asis of a small number of samples so far available, to be some-
nat richer in sucrose, as are those visited by non-volant
ammals.

    Many of the Microchiroptera that visit flowers are also
ruit juice swallowers and may have picked up a desire for
exoses from the juices. However, the Megachiroptera are also
ruit juice swallowers so that cannot be the whole story.

    3) Lepidopteran-pollinated flowers -- Crepuscular and
octurnal visits to flowers by swift-flying and hovering hawk-
oths (Sphingidae) as well as those by the less-dramatic
settling" moths (Noctuidae, Geometridae, Pyralidae, etc.)

---

able 6. Mean Sugar Ratios ($\frac{S}{G+F}$) in Erythrina *spp.*

ummingbird Pollinated	*Passerine bird Pollinated*	
	Old World Species	*New World Species*
$\bar{x} = 1.135$	$\bar{x} = 0.046$	$\bar{x} = 0.033$
(N = 21)	*(N = 11)*	*(N = 8)*
ange 0.557 - 2.187	*Range 0.017 - 0.074*	*Range 0.015 - 0.049*

are rewarded with nectars that are characteristically sucrose-
rich or sucrose-dominant (table 5)(Baker, H.G., 1978; Baker,
H.G. and I. Baker, 1979a, 1981), in strong contrast to the
hexose-richness of the nectars of New World bat-flowers that
are equally nocturnal in anthesis. Again, the nature of the
regulating physiological phenomena in the moths which leads
to this pattern of nectar-sugar provision is not yet known.
It is tempting to see in this contrast an isolating barrier
between flowers whose openings are temporally coincident but
are adapted to different pollinators. For this connection,
the two cases we are aware of, and have photographed, of moths
visiting "bat-flowers" have both been to trees from the Old
World, *i.e., Durio zibethinus* (Bombacaceae)(sugar ratio 0.754)
(Baker, H.G., 1970) and *Kigelia africana* (Bignoniaceae)(Harris
and Baker, 1958).

"Butterfly-flowers" with long corolla tubes are diurnal
in flowering and they usually produce nectar that is (like
that of the moth-flowers) sucrose-rich or dominant (table 5).
However, butterflies are also involved with another class of
flowers that we have called "bee- and butterfly-flowers."
These are generally short-tubed and include many Asteraceae
(Compositae) where the capitulum forms a suitable "standing
platform" for the insects. These flowers tend to have hexose-
rich or hexose-dominant nectar (table 5)(Baker, H.G. and I.

FIGURE 1. Glossophaga soricina *(Microchiroptera) visiting
pendulous inflorescences of the vine* Mucuna
andreana *(Fabaceae) at Santa Ana, Costa Rica.*

FIGURE 2. A *"flying fox",* Pteropus tonganus *(Megachiroptera), pauses between nectar collecting
episodes from* Ceiba pentandra *(Bombacaceae)
near Nadi, Fiji.*

Baker, 1981). In this, they agree with the flowers of the
"short-tongued bee" class, for the bees that visit them are
generally of the short-tongued types.

  4) Hymenopteran-pollinated flowers -- Nectars from "bee-
flowers" as a whole are a mixed bunch but the situation is
clearer if they are divided into two groups -- "long-tongued
bee-flowers" (whose visitors have tongue-lengths greater than
6 mm) and "short-tongued bee-flowers" (whose visitors' tongues
are less than 6 mm in length). Clearly, long-tongued bees
appear to be rewarded usually with sucrose-rich or sucrose-
dominant nectar (table 5)(Baker, H.G. and I. Baker, 1981).
On the other hand, short-tongued bees (and, incidentally, the
flies that often visit the same open or short-tubed flowers)
imbibe nectar that is usually hexose-rich or dominant.

  "Wasp-flower" nectars (which are more common in the tropics
than in the temperate regions) usually seem to be rather rich
in sucrose (table 5).

  5) Individual flower genera and inheritance -- The example
of hummingbird-pollinated and passerine bird-pollinated
Erythrina species shows that evolutionary change (in this
case probably from hexose-domination to sucrose-richness or
dominance when the genus came into contact with hummingbirds
in the New World) can take place within a single genus, even
differentiating species that are phylogenetically still close
enough to hybridize. In *Erythrina* two horticulturally made
hybrids provide information about the inheritance of the
sugar-ratio characteristics (table 7). In the hybrid between
E. herbacea (sucrose-rich, 0.668) and *E. crista-galli* (hexose-
dominated, 0.034) the nectar sugar-ratio is intermediate (0.217)
(Baker, I. and H.G. Baker, 1979). In the case of *E. resupinata*
(0.030) x *E. variegata* (0.017), the parents both have hexose-
dominated nectar, so it is not surprising that this also

Table 7. Sugar proportions and ratios of species pairs and their hybrids.

	N	Melez.	Malt.	Sucr.	Gluc.	Fruct.	$\frac{S}{G+F}$
Erythrina herbacea	2	.017	N.D.	.394	.318	.272	0.668
E. x bidwilli	2	.008	N.D.	.177	.434	.381	0.217
E. crista-galli	5	.009	.015	.032	.417	.527	0.034
Erythrina resupinata	1	N.D.	.029	.029	.639	.302	0.030
E. x resuparcelli	2	N.D.	.029	.044	.532	.395	0.047
E. variegata var. parcelli	3	.004	.006	.008	.529	.453	0.017
Penstemon centranthifolius	3	.015	N.D.	.571	.234	.180	1.401
P. x parishii	2	.025	N.D.	.545	.289	.141	1.271
P. spectabilis	3	.006	.005	.206	.483	.300	0.273
Campsis radicans	4	.015	.012	.616	.152	.206	1.721
C. x tagliabuana	10	.006	.011	.054	.578	.351	0.058
C. grandiflora*	?	N.D.	N.D.	trace	.600	.400	near 0

*From Colliva and Giulini (1970)

N.D. = not detected

characterizes the hybrid (0.047)(Baker, H.G. and I. Baker, in prep).

In *Penstemon* (Scrophulariaceae)(along with *Chionophila* and *Keckiella*), 7 hummingbird-pollinated species have a mean sugar ratio of 1.075 (with a range from 0.502 to 1.401), while 17 species are pollinated by various Hymenoptera and show a mean ratio of 0.300 (with a range from 0.045 to 0.853). *P. centran-thifolius* (hummingbird, 1.401) hybridizes with *P. spectabilis* (wasp, 0.274)(table 7). In this case, the hybrid is *P. x parishii* (sugar ratio, 1.271) and closely resembles the sucrose-dominant parent.

In *Campsis* (Bignoniaceae), horticultural hybrids have been made between *C. radicans* (hummingbird) and *C. grandiflora* (sunbirds). The hybrids (by now a swarm) show ratios that are closer to that of the hexose-dominant parent.

Thus, there is evidence that the genetics of sugar-ratio determination are different for different taxa.

The *Campsis, Penstemon* and *Erythrina* examples show how apparently adaptive changes in sugar ratios can occur with the evolution of related species. However, there are some other pieces of evidence that "phylogenetic constraint" may sometimes prevent full adaptation to a new kind of pollinator. Thus, *Mutisia viciaefolia* (Asteraceae), from the high Andes of Peru, has a full syndrome of adaptations to pollination by hummingbirds (being unusual in the Asteraceae in utilizing this kind of pollinator) except for its hexose-rich nectar (the common type in the Asteraceae)(Baker, H.G. and I. Baker, 1981). The case of *Mutisia* and its apparent "phylogenetic constraint" reminds us that some angiosperm *families* are characterized by particular patterns of nectar sugar proportions, so that the ratios, which must have a strong genetic basis, are conservative characters in evolution.

AMINO ACIDS

A history of the determination of the presence of amino
acids in nectar is provided in Baker, H.G. and I. Baker (1975).
The first person to recognize their presence seems to have
been Ziegler (1956). Since we began the study of nectar amino
acids in 1972, we have published surveys (Baker, H.G. and I.
Baker, 1973, 1975; Baker, H.G., 1977, 1978); evidence of
consistency of amino acid complements in the face of environmental and genetical variation (Baker, H.G. and I. Baker,
1977); evidences of inheritance in species-hybrids and
usefulness in establishing phylogeny (Baker, I. and H.G. Baker,
1976, 1979; Baker, H.G. and I. Baker, 1977); and comparisons
of floral and extrafloral nectars (Baker *et al.*, 1978). Our
methodology is described (with improvements in the later
papers) in Baker, H.G. and I. Baker, (1973, 1975), Baker, I.
and H.G. Baker, (1976, 1979) and Baker, *et al.*, 1978). These,
and the many contributions of other workers to an understanding
of the occurrence and biological significance of amino acids
in floral nectar, will be listed and reviewed in another
publication. Here, we may simply point to conclusions
relevant to pollination biology and phylogeny.

Concentrations of Amino Acids

In our surveys, attention was given first to the overall
concentrations of amino acids in floral nectars. Such
concentrations are estimated by staining dried spots of
nectar on chromatography paper with ninhydrin in solution in
acetone. The color that develops may be compared with that
produced by staining spots of known concentrations of
histidine (Baker, H.G. and I. Baker, 1973, 1975, etc.) or,
when sufficient material is available, the color is eluted with

50% aqueous methanol and the concentration read on a
spectrophotometer at 570 nm (see Baker, I. and H.G. Baker,
1981, for details).

A general correlation of increased amino acid concentration with seven morphological features of flowers generally
conceded to be "advancement" indicators was demonstrated early
on (Baker, H.G. and I. Baker, 1973, 1975). Presumably, this
can be interpreted as an increase in amino acid concentrations
in those flowers where the nectar has to be sought rather than
being exposed to all comers. However, there appears also to
be a tendency in contemporary angiosperms of many families for
the average concentrations in nectars from woody plants to be
lower than those in the flowers of herbaceous plants in the
same family or in the same community, and even some evidence
from tropical woody plants that amino acid concentrations are
lower in flowers borne high on the tree compared to those
borne on the lower branches (Baker, H.G., 1978).

Table 8 shows some of our data relating nectar amino
acid concentrations to pollinator types. Determinations were
made at Berkeley from nectar-collections made in various
temperate and tropical regions.

Heading the list, with extremely strong amino acid
concentrations, are the flowers that simulate carrion or
dung and attract the females of flies that specialize in
laying their eggs in such media. These flowers may look and
smell like these animal materials and it is appropriate that
they should produce nectar with a high amino-acid content --
where they produce nectar at all; for, in the Asclepiadaceae,
they include some of the most arrant deceivers, with shiny
surfaces on the throats of the corollas that are not producing
liquid at all (Baker, H.G. and I. Baker, 1975). By contrast,
nectar taken by syrphids, bombyliids, and muscids, and less-
specialized flies, mostly come from open bowl or short-tubed

flowers and show relatively low levels of amino acids (table 8).

Our results show that the concentration of amino acids in nectar is greater if nectar is the only or predominant source of protein-building material for the adult visitor or its brood. It is strikingly lower if the visitors make much use of pollen, or, in the case of vertebrate visitors, insects as alternative foods. Thus, in table 8, it may be seen that the nectars of flowers visited by butterflies, settling moths and wasps, respectively, all show high levels of amino acid concentration. By contrast, the figures for the nectars of bee- and fly-flowers are relatively low. The separated class of "bee- and butterfly" - flowers produce nectars that are more like those of the butterfly flowers (although it will be remembered that, in sugar ratios, they are more like those

Table 8.   Average amino acid concentrations in floral nectars, grouped according to principal type of pollinator. Determinations made at Berkeley from nectar collections in various temperate and tropical regions.

Principal Pollinator	Number of determinations	Amino acids in micromoles per ml.
Carrion & dung flies	9	12.500
Butterflies	118	1.148
Settling moths	78	1.059
Bees + Butterflies	257	1.015
Wasps	44	0.913
Bees	715	0.624
Flies (general)	97	0.557
Hawkmoths	65	0.536
Hummingbirds[1]	150	0.452
Bats	23	0.306
Passerine birds[2]	21	0.255

[1] excludes Erythrina hummingbird-pollinated nectars

[2] excludes Erythrina passerine bird-pollinated nectars

(see Table 9)

of short-tongued bee-flowers).

Hawkmoths (Sphingidae) imbibe large quantities of nectar each night and the amino acid concentration in these nectars is low (table 8); nevertheless, the total amount of amino acids ingested may be high.

A feature of lepidopteran-pollinated flowers that have a narrow corolla-tube containing introrsely-dehiscing anthers is the probability that the concentration of amino acids in the nectar at the base of the tube will be significantly increased by dislodged pollen falling in, especially after the probing mouth parts of an insect have been in action (Baker, H.G. and I. Baker, 1975), a point that has been reiterated recently by Willmer (1980).

Flower-visiting bats (of the sub-order Microchiroptera) in the New World tropics and subtropics may not catch many insects, but they make good use of pollen as a protein source (Howell, 1974) and are usually partakers of fruit juices, as well. Consequently, it is not surprising that bat-flower nectars are weak in amino acids (table 8)(see also Voss *et al.*, 1980).

Among bird-visitors to flowers, the hummingbirds are insect-catchers, and floral nectars could hardly be expected to provide significant competition with the insects as sources of protein-building amino acids, and they do not (table 8).

Nectarivorous passerine birds may also be insectivorous, as well as taking fleshy and juicy fruits in tropical and subtropical ecosystems, and, as a rule, they find low concentrations of amino acids in the nectars of the flowers they visit. However, in the tropical tree genus *Erythrina* (Fabaceae), where there are some species primarily visited by nectarivorous passerine birds, while the remainder have predominantly hummingbird-visited flowers, the nectar amino

acid concentrations of the former are significantly higher
than those of the latter (Table 9)(Baker, I. and H.G. Baker,
1981 -- and see, also, Cruden and Toledo, 1977). This may be
related to a slow-down in insect-catching and fruit-taking
at a time when the *Erythrina* trees are massively in flower
and at least the larger passerine birds are defending
territories at the trees (Cruden and Toledo, 1977; Cruden and
Hermann-Parker, 1977). The nectar of *E. breviflora* was the
first in our experience to contain all ten of the "essential
acids" for birds (see Cruden and Toledo, 1977).

The conclusion that can be drawn from these data is that
for Lepidoptera and wasps, the amino acids are in adequate
quantities to perform a nutritive function (Baker, H.G. and
I. Baker, 1975). In this connection, it may be noted that
feeding experiments with *Colias* butterflies (Watt, *et al.*,
1974) and with butterflies of several species (Arms *et al.*,
1974), as well as the natural reinforcement of nectar amino
acids by those diffusing out of "dunked" pollen by *Heliconius*
species (Gilbert, 1972; Dunlap-Pianka *et al.*, 1977) and
possibly other butterflies (DeVries, 1979), all suggest that
amino acids taken in in nectar may positively aid in repair
of protein-breakdown, increase longevity of the imago, and
increase egg-output by females in those species where egg-
production is not complete at the time of emergence of the
imago from the pupa.

Table 9. Mean amino acid concentrations Erythrina *spp.*
        (in μmol/ml.*)*

Hummingbird Pollinated	Passerine bird Pollinated	
	Old World Species	New World Species
$\bar{x} = 2.78$	$\bar{x} = 28.06$	$\bar{x} = 24.64$
(N = 21)	(N = 11)	(N = 8)

Even in those cases where the amino acid concentration is too low for them to have direct nutritive significance, it is likely that their presence, proportions and concentrations may influence the "taste" of the nectar (Baker, H.G. 1977, 1978) and, along with the sugars, may reinforce the morphology, color and scent of flowers in enabling them to build a relationship with a particular pollinator species.

Studies of amino acid concentrations in the nectars of plants growing in the alpine and subalpine zones of the Rocky Mountains, have shown (Table 10) that they are unusually highly concentrated for each of the pollinator classes represented there (Baker, H.G. and I. Baker, in prep.). Thus, it must be recognized that either directly or indirectly through selection, an extreme environment may have a significant effect on nectar amino acid concentration and this must be watched for in other situations.

Complements of Amino Acids

In our analyses of the amino acid complements of nectars we first made use of paper chromatography (Baker, H.G. and I. Baker, 1973, 1975), but the sensitivity of this technique

Table 10. *Comparison of all amino acid concentrations with those of Colorado alpine and subalpine plants in selected pollinator groups.*

	Bee (N)	Bee + Butterfly (N)	Fly (N)	Humming- bird (N)
Overall	0.624 (715)	1.015 (257)	0.557 (97)	0.432 (150)
Alpine + Subalpine	2.012 (87)	3.476 (44)	2.543 (37)	1.825 (6)

Concentrations in μmol/ml.

was less than satisfactory. Consequently, we turned to
polyamide thin-layer-chromatography on a miniaturized scale
(Baker, I. and H.G. Baker, 1976; Baker et al., 1978). We
did not use an automatic amino acid analyzer (which was not
easily available and would have been too expensive to
operate with the numbers of species that we have been examining)
because we often have only a few microliters of nectar available
The polyamide T.L.C. of the dansylated amino acids, with
identification of and quantification by utilization of the
U.V. fluorescence, is described in Baker, H.G. and I. Baker
(1977) and Baker, et al., (1978). The results show remarkable
consistency in multiple determinations from separate plants
under varying environmental conditions (Baker, H.G. and I.
Baker, 1977; Baker, H.G., 1978). Care is always taken that
no pollen has fallen into freshly produced nectar.

A summary of the proportions of nectars containing the
various "protein-building" amino acids for 395 floral nectars
(using the most up-to-date of our methods), arranged in order
of frequency of occurrences, is given in table 11 (data from
Baker, H.G., 1978, with additions). Although there is wide
difference between the frequency of occurrences of individual
amino acids, it is clear that all can be available if a range
of nectars is collected by a flower-visitor.

Inheritance of the amino acid complements in crosses
between species appears to be additive (on a presence or
absence basis) in the F_1 generation (Baker, I. and H.G. Baker,
1976, with examples from Aloë, Oxalis, Geranium, Silene,
Cercidium and Armeria; Baker, H.G. and I. Baker, 1977, from
Brassica, and Baker, I. and H.G. Baker, 1979, from Erythrina).
Table 12 repeats the picture for Erythrina herbacea crossed
with E. crista-galli and gives new data for E. resupinata
crossed with E. variegata. Again additive qualitative

Table 11. Frequencies of occurrence of individual amino acids
           in floral nectars of 395 species.

	Detected in	Proportion
Alanine	380	.96
Arginine	356	.90
Serine	352	.89
Proline	344	.87
Glycine	332	.84
Isoleucine	287	.73
Threonine	263	.67
Valine	260	.66
Leucine	255	.65
Glutamic	245	.62
Cysteine, etc.	218	.55
Phenylalanine	216	.55
Tyrosine	204	.52
Tryptophan	189	.48
Lysine	162	.41
Glutamine	162	.41
Aspartic	128	.32
Asparagine	106	.27
Methionine	80	.20
Histidine	77	.19
Non-protein	144	.36

inheritance is shown.

Hybrid swarms in *Aquilegia* and *Penstemon* show recom-
binations in what must be hybrid generations after the first
and backcrosses (Baker, I. and H.G. Baker, 1976). The
utility of nectar amino acids, with the additive inheritance,
in resolving problems of allopolyploidy versus autopolyploidy
has been pointed out (Baker, I. and H.G. Baker, 1976; Baker,
H.G. and I. Baker, 1977).

However, the similarity between the constituents of nectars
from closely related species suggests a high degree of
"phylogenetic constraint." This is illustrated by *Penstemon*
species in subgenera Penstemon, Saccanthera, Dasanthera and
in *Keckiella* (= subg. Hesperothamnus). This (table 13) is
the result of work done with a less satisfactory polyamide

Table 12. Amino acid complements of two pairs of Erythrina species and their hybrids. The proportions of each acid in the total for each taxon are shown.

Amino acid	Erythrina crista-galli	Erythrina x bidwillii	Erythrina herbacea	Erythrina variegata	Erythrina x resuparcelli	Erythrina resupinata
Alanine	.101	.089	.039	.087	.054	.020
Arginine	.069	.028	.004	.008	.031	.037
Asparagine	.065	.111	.174	.193	.157	.146
Aspartic	.006	.024	.036	.008	.017	.008
Cysteine, etc.	.003	.019	.018	.009	.025	.025
Glutamic	.006	.047	.038	.006	.009	.004
Glutamine	.202	.177	.231	.053	.051	.039
Glycine	.016	.029	.054	.025	.020	.016
Histidine	.012	.008	.025	.035	.035	N.D.[a]
Isoleucine	.065	.044	.043	.037	.026	.033
Leucine	.017	.017	.036	.043	.025	.031
Lysine	.025	.028	.032	.036	.023	.029
Methionine	.014	.018	N.D.[a]	.028	.038	.033
Phenylalanine	.023	.039	.027	.050	.029	.047
Proline	.075	.075	.076	.181	.112	.082
Serine	.074	.058	.028	.034	.039	.041
Threonine	.071	.041	.024	.016	.035	.037
Tryptophan	.010	.022	.029	.018	.021	.049
Tyrosine	.019	.037	.009	N.D.[a]	.058	.058
Valine	.098	.067	.058	.053	.053	.061
Unknown #1	.013	.006	N.D.[a]			
Unknown #2	N.D.[a]	.018	.022			
Unknown #3				N.D.[a]	.137	.186
γ-amino-butyric				.078	.011	.018

[a]N.D. = not detected

Table 13. *Amino acid complements of species in three subgenera of* Penstemon *and the segregate genus* Keckiella *(=* Penstemon *subg.* Hesperothamnus*)*

	Subg. Penstemon							Subg. Saccanthera					Subg. Dasanthera		Keckiella (Subg. Hesperothamnus)			
	heterodoxus	deustus	spectabilis	centranthifolius	oreochoris	procerus	speciosus	gracilentus	azureus	hetophyllus	bridgesii	laetus	davidsonii	newberryi	breviflorus	cordifolius	antirrhinoides	ternatus
Alanine	+	+	+	+	+	+	+	+	+	+	+	+	+	+	+	+	+	+
Arginine	+	+	+	+	+	+	+	+	+	+	+	+	+	+	+	+	+	+
Asparagine																		
Aspartic																		
Cysteine, etc.	+						+	+							+			
Glutamic	+	+	+	+	+		+	+		+			+					+
Glutamine																		
Glycine	+	+	+	+	+	+	+	+	+				+					+
Histidine		+																
Isoleucine	+	+	+	+	+	+	+	+										
Leucine	+	+	+	+	+	+	+											
Lysine	+	+	+	+	+													
Methionine																		
Phenylalanine																		
Proline	+	+	+	+	+	+	+				+		+					
Serine	+	+	+	+	+	+	+	+	+	+	+	+	+	+	+	+	+	+
Threonine	+	+	+	+	+	+	+	+	+	+	+	+	+	+	+	+	+	+
Tryptophan					+	+				+								
Tyrosine	+	+						+	+									
Valine	+	+	+	+		+				+								
Unknown #1		+																
Unknown #2	+																	
Total	10	13	13	14	12	10	12	5	9	5	6	4	5	6	5	4	4	6

sheet than we now use, and there may have been loss of distinction between amino acids, especially between leucine and phenylalanine, but the structure of the picture is clear; close

resemblance between the nectar amino acid complements within
subgenera, but substantial differences between the subgenera.
All three subgenera and *Keckiella* contain some species
pollinated by bees and others pollinated by hummingbirds, yet
within each subgenus the species show very similar complements;
only the concentrations of the amino acids may vary with
pollinator. In this connection we should remember that the
nectar sugar-ratios did vary with pollinator type in *Penstemon.*

   In the case of *Erythrina herbacea* and *E. crista-galli,*
the nectar sugar-ratios are very different (0.668 and 0.034,
respectively), and we have suggested (Baker, I. and H.G.
Baker, 1979) that *E. crista-galli* is pollinated primarily by
perching birds. However, once again the amino acid complements
are very similar (differing only in the absence of methionine
and one "unknown" in *E. herbacea* and another "unknown" in
E. crista-galli)(table 12).

NON-PROTEIN AMINO ACIDS

   The presence in many complements of unusual amino acids,
beyond those that are known to be involved in protein-building,
with a higher proportion of tropical than temperate plants being
positive for one or more of them (Baker, H.G., 1978), merits
consideration, but until more of them can be identified their
biological significance will not be known. Some non-protein
amino acids in seeds have been regarded as chemical defenses
against seed-predators (Rehr, *et al.,* 1973). It is possible
that they may perform a similar role in nectar, although it
must be remembered that the function of nectar is the reward
of pollinators and not their poisoning (see also Rhoades and
Bergdahl, 1981). Also, it is possible that some of these
amino acids, such as γ-amino butyric acid, are not very toxic
and may serve some other function than chemical defense. When

they are identified they may be as useful taxonomically and
phylogenetically as the non-protein amino acids in vegetative
parts and seeds.

LIPIDS

The first report of lipids as liquid rewards to pollinators
visiting flowers was made by Vogel (1969), with subsequent
papers enlarging the picture (Vogel, 1971, 1974, 1976). Except
for the last paper, which deals with *Lysimachia* (Primulaceae)
in Europe, the references are to certain South American species
of Malpighiaceae, Iridaceae (*e.g., Sisyrinchium*), Scrophulariaceae (*e.g., Calceolaria),* Krameriaceae (*Krameria*) and
Orchidaceae (*e.g., Oncidium*). Vogel's observations implicated
several anthophorid bees, but most obviously species of
Tapinotaspis and *Centris*, which collect these lipids and use
them, along with pollen, to provision the nest cells containing
larvae. Vogel asserted (1969, 1971, 1974) that the production
of lipids by glands that he named "elaiophors" is an *alternative*
to the production of aqueous nectar. Subsequent studies of
elaiophors and their significance in pollination have been
made by Buchmann (pers. comm.) on *Mouriri* (Melastomataceae)
and Simpson *et al.* (1977) on *Krameria*.

However, being alerted by Vogel's work to the possibility
of lipids as rewards to flower-visitors, we soon found that
many nectars contain lipids (Baker, H.G. and I. Baker, 1973,
1975; Baker, H.G., 1977, 1978). Our method was a simple one
that may not pick up all lipids (black color of the sample
spots on chromatography paper with osmic acid, with a follow-
up of fresh nectar, if this is available, stained with Sudan
III). We found that approximately one-third of all species
tested showed lipid presence in amounts ranging from a trace
to quantities sufficient to make the nectar milky in appearance

(*e.g., Catalpa speciosa*). Consideration of data available in
1976 (Baker, H.G., 1978) suggested that, in tropical forests
in Costa Rica, at least, lipids in nectar occur more frequently
in trees than in any other life form.

"ORGANIC ACIDS"

Nectars that contain lipids frequently contain antioxidant
organic acids (Baker, H.G. and I. Baker, 1975), notably ascorbic
acid (Vitamin C), which may play a rancidity-preventing role.
Whether or not the organic acids are the main causes of acidity
in nectar, some nectars are strongly acid (reaching pH 3.0 in
Silene alba, Baker, H.G. and I. Baker, unpubl.). However,
others are alkaline in reaction (with an extreme, in our
experience, of pH 10.0 in *Viburnum costaricanum*). The
contribution of the various chemical components of nectar to
the pH, and the pollinatory or phylogenetic significance of
their variations will be the subject of further research in
our laboratory. Possibly inorganic substances may contribute
to the pH picture, and we must examine them.

PROTEINS

As nutritional items, the proteins in nectar do not appear
to be significant. Some nectars contain none that can be
detected by the brom-phenol blue test. This is in striking
contrast to pollen, where the protein content is high (built
up from a wide range of amino acids). Presumably, the
proteins in nectar are largely enzymatic. Invertase, trans-
glucosidase, phosphatases, oxidases and tyrosinase have been
reported to occur in nectars from various species (see listing
in Baker, H.G. and I. Baker, 1975, p. 117). More recently,
Scogin (1979) demonstrated the presence of esterases and malate

dehydrogenase in nectar of *Fremontia* (Sterculiaceae). It is
unclear how often the interconversion of sucrose and hexose
sugars takes place in nectar under the influence of enzymes
contained therein, but it is virtually certain that each of
these sugars can be excreted in final form from the nectary,
and the presence of all three together in nectar does not by
itself imply enzymatic activity after excretion.

NON-NUTRITIVE SUBSTANCES

Alkaloids

This heterogeneous group of substances is included in the
list of nectar-constituents on the basis of spot tests with
iodoplatinate and Dragendorff reagents on dried nectar samples.
By 1977 (Baker, H.G., 1977, 1978), it was possible to report
that 50 out of 593 tests were positive and it was pointed
out that the proportion of nectars containing alkaloids was
higher in the tropics than in lowland California and very low
in the alpine tundra of the Colorado Rockies. Data obtained
subsequently have given confirmatory evidence presented in
table 14.

Alkaloids, in addition to a possible function as poisonous
substances in plant tissues, may also be temporary storage
substances for nitrogen (Robinson, 1968). However, it does
not seem likely that this would apply to nectar, which is
probably not going to be resorbed into the plant.

Phenolics

Another very heterogeneous group of chemicals is the
phenolics, whose presence in nectar spot samples is detected

Table 14. Frequences of occurrence of nectars containing alkaloids and phenolics respectively,
in plants of all life forms on a transect from tropical forests in Costa Rica to the
alpine tundra of Colorado, U.S.A.

	Alkaloids		Phenolics	
	number of spp. examined	% with alkaloids	number of spp. examined	% with phenolics
Costa Rica – lowland forest	266	12.0	263	49.0
Costa Rica – cloud forest	163	9.8	186	45.2
California – below 2,438 m.	275	6.9	248	29.6
California – above 2,438 m.	67	7.4	54	29.0
Colorado – subalpine forest	62	4.8	78	32.0
Colorado – alpine tundra	77	0	77	32.0

by color-testing with p-nitraniline. To date, only Scogin
(1979) has attempted an identification: an isoflavone, 5,7-
dimethoxygenistein-4'-glucoside, in *Fremontia*. In 1977
(Baker, H.G., 1977, 1978), we had evidence of phenolics in
191 out of 528 species tested (36.2%). Again there appeared
to be a larger proportion of "positives" in tropical rather
than temperate regions of California, and a still lower
proportion in the subalpine and alpine zones in the latter
state and Colorado. This is substantiated by the more recent
studies of the occurrences of these substances (table 14), in
a survey that now covers 850 species.

Ant repellents

A suggestion by D.H. Janzen (1977) that ants rarely take
floral nectar even when this is freely exposed by tropical
plants, and that this might be due to the frequent presence
of deterrent chemicals, produced a flood of replies by other
workers in the pages of 'Biotropica'. The consensus seems to
be that there may be a small proportion of floral nectars
that are chemically protected from ants. *Hippobroma longiflora*
(Campanulaceae) has an ant repellent chemical (Feinsinger and
Swarm, 1978, but W.A. Haber (pers. comm.), and E.O. Guerrant
and P. Fiedler (1981), have evidence that the toxic substance
is in the petals and is released into the nectar when the
corolla-tube is damaged by the probing mouthparts of insects.
The morphology of flowers may be more important than chemical
defenses in keeping ants out of them.

PHYLOGENETIC CONSTRAINTS

Most particularly in the sugar and amino acid concentrations
of nectars, it is apparent that, although natural selection by

pollinators can have a strong effect on their composition,
there appear to be constraints imposed by the limited variety
of the genetic material that has been handed down to contemporar
plants from their ancestors. This "phylogenetic constraint" may
limit the range of pollinators that may be utilized by the
flowers concerned, or it may be that a relationship with a
pollinator is struck up with less than 100% of the chemical
adaptation in some cases. It is important to recognize that
adaptation does not have to be 100% perfect as long as potential
competitors for the services of the pollinator are still farther
from perfection. Indications of "phylogenetic constraint" are
also noticeable in the pollen studies that we are also making
(Baker, H.G. and I. Baker, 1979b).

CONCLUSIONS

In these discussions of nectar chemistry there is clear
evidence of adaptation by flowering plants to the reward of
flower-visitors in the operation of pollinatory mechanisms.
There is also evidence, not considered here, of reciprocal
adaptation by the pollinator to the flowers. These processes
are usually referred to as 'co-evolution' [but might better be
called 'reciprocating evolution' (Baker, H.G. and P.D. Hurd)] .
Thus, in addition to generally appropriate sugar concentrations, the sugar ratios of nectars are remarkably consistent
within a species and are generally of the same order for
related species that are rewarding a particular category of
flower-visitor. Amino acid concentrations also appear to be
generally in harmony with the needs of the visitors for
protein-building materials derived from nectar.
Nevertheless, there are also evidences of 'phylogenetic
constraint' or adaptation to be seen in the sugar ratios and,
even more obviously, in the amino acid complements of nectar.

The near uniformity of sugar ratio classes in some large
families and the similarity of amino acid complements in
related species suggest close genetic control of production
and secretion of these chemicals, and this genetic control
may be conservative. Thus, there is an interplay between
adaptation and conservatism which is a familiar feature of
evolutionary biology. These limitations to less than 100%
of the ideal syndromes of characters need not be fatal to the
success of the pollinatory interaction provided that potential
competitors for the services of the pollinator do not reach
a higher level of adaptation.

   Most pollinator-reward studies have been reported in terms
of individual species and their nectar chemistry. Clearly
there is a need to put all biological studies of this sort
on an ecosystem basis rather than merely considering
individual taxa as though they evolved *in vacuo*. The
differences in overall sugar concentrations and in overall
amino acid concentrations that we have reported between
tropical and high montane temperate samples indicate the
importance of the habitat in determining the amounts of the
nectar rewards while retaining the characteristic pollinator-
adaptation. Our reference to the proportions of nectars
containing alkaloids and phenolics on these regional bases is
another step in this direction which we are following with
others.

ACKNOWLEDGEMENTS

   We are indebted to a number of biologist colleagues who
have supplied us with nectar samples from various parts of
the world. Also, C. Turner, D. Schoen, E.O. Guerrant and
F. Bowcutt have shared in recent laboratory work. Colleagues

we have worked with in the field include G.W. Frankie, P.A.
Opler, J.H. Bock, R.W. Cruden, P.G. Kevan, K.H. Keeler,
S.C.H. Barrett, S. Koptur, P.F. Yorks and P. Burton. All
of these persons and many besides have stimulated and
informed us in discussions. The National Science Foundation
assisted us immeasurably with financial support. We extend
our thanks to all.

LITERATURE CITED

ARMS, K., P. FEENY, and R.C. LEDERHOUSE. 1974. Sodium:
    Stimulus for puddling behavior by Tiger Swallowtail butter-
    flies. Papilio glaucus. Science 185:372-374.
BAKER, H.G. 1970. Two cases of bat pollination in Central
    America. Rev. Biol. Trop. 17:187-197.
BAKER, H.G. 1975. Sugar concentrations in nectars from
    hummingbird flowers. Biotropica 7:37-41.
BAKER, H.G. 1977. Non-sugar chemical constituents of nectar.
    Apidologie 8:349-356.
BAKER, H.G. 1978. Chemical aspects of the pollination biology
    of woody plants in the tropics. Chap. 3, Pp. 57-83. *In:*
    P.B. Tomlinson and M.H. Zimmermann, eds. Tropical Trees
    as Living Systems. Cambridge: Cambridge University Press.
BAKER, H.G., and I. BAKER. 1973. Some anthecological aspects
    of the evolution of nectar-producing flowers, particularly
    amino acid production. Chap. 12, Pp. 243-264. *In:* V.H.
    Heywood, ed. Taxonomy and Ecology. London: Academic
    Press.
BAKER, H.G., and I. BAKER. 1975. Studies of nectar-constituti
    and pollinator-plant coevolution. Pp. 100-140. *In:* L.E.
    Gilbert and P.H. Raven, eds. Coevolution of Animals and
    Plants. Austin: University of Texas Press.
BAKER, H.G., and I. BAKER. 1977. Intraspecific constancy of
    floral nectar amino acid complements. Bot. Gaz. 138:183-19
BAKER, H.G., and I. BAKER. 1979a. Sugar ratios in nectars.
    Phytochem. Bull. 12:43-45.
BAKER, H.G., and I. BAKER. 1979b. Starch in angiosperm
    pollen grains and its evolutionary significance. Amer. J.
    Bot. 66:591-600.
BAKER, H.G., and I. BAKER. 1981. Floral nectar sugar con-
    stituents in relation to pollinator type. In press, *In:*
    C.E. Jones and R.J. Little, eds. Handbook of Experimental
    Pollination Biology. New York: van Nostrand-Reinhold.

BAKER, H.G., and P.D. HURD. 1968. Intrafloral ecology. Annu. Rev. Entom. 13:385-414.

BAKER, H.G., P.A. OPLER, and I. BAKER. 1978. A comparison of the amino acid complements of floral and extrafloral nectars. Bot. Gaz. 139:322-332.

BAKER, I., and H.G. BAKER. 1976. Analyses of amino acids in floral nectars of hybrids and their parents, with phylogenetic implications. New Phytol. 76:87-98.

BAKER, I., and H.G. BAKER. 1979. Chemical constituents of the nectars of two *Erythrina* species and their hybrid. Ann. Missouri Bot. Gard. 66:446-450.

BAKER, I., and H.G. BAKER. 1981. Some chemical constituents of floral nectars of *Erythrina* in relation to pollinators and systematics. Allertonia (in press).

COLLIVA, V., and F. GIULINI. 1970. Osservasioni sull'ibrido *Campsis xtagliabuana* (Vis.) Massalongo e sulle sue specie genitrice. Giorn. Bot. Ital. 104:469-482.

CORBET, S.A. 1978. Bees and the nectar of *Echium vulgare*. Pp. 21-29. *In:* A.J. Richards, ed. The Pollination of Flowers by Insects. London: Academic Press.

CRUDEN, R.W., and S. HERMANN-PARKER. 1977. Defense of feeding sites by orioles and hepatic tanagers in Mexico. Auk 94:594-596.

CRUDEN, R.W., and V.M. TOLEDO. 1977. Oriole pollination of *Erythrina breviflora* (Leguminosae): Evidence for a polytopic view of ornithophily. Pl. Syst. Evol. 126:393-403.

DeVRIES, P.J. 1979. Pollen-feeding rainforest *Parides* and *Battus* butterflies in Costa Rica. Biotropica 11:237-238.

DUNLAP-PIANKA, H., C.L. BOGGS, and L.E. GILBERT. 1977. Ovarian dynamics in heliconiine butterflies: Programmed senescence versus eternal youth. Science 197:487-490.

FEINSINGER, P., and L.A. SWARM. 1978. How common are ant-repellent nectars? Biotropica 10:238-239.

GILBERT, L.E. 1972. Pollen feeding and reproductive biology of *Heliconius* butterflies. Proc. Nat. Acad. Sci. U.S.A. 69:1403-1407.

GRANT, K.A., and V. GRANT. 1968. Hummingbirds and Their Flowers. New York: Columbia University Press.

GUERRANT, E.O., and P.L. FIEDLER. 1981. Flower defenses against nectar-pilferage by ants. Biotropica 13(2): Supplement 25-33.

HAINSWORTH, F.R., and L.L. WOLF. 1976. Nectar characteristics and food selection by hummingbirds. Oecologia (Berl.) 25:101-114.

HARBORNE, J.B. 1977. Introduction to Ecological Biochemistry. London: Academic Press.

HARRIS, B.J., and H.G. BAKER. 1958. Pollination in *Kigelia africana* (Benth.). J. West Afr. Sci. Ass. 4:25-30.

HAYDAK, M.H. 1979. Honey bee nutrition. Annu. Rev. Entomol. 15:143-156.

HOCKING, B. 1953. The intrinsic range and speed of flight
    of insects. Trans. Roy. Ent. Soc. Lond. 104:223-345.
HOWELL, D.J. 1974. Bats and pollen: Physiological aspects
    of the syndrome of chiropterophily. Comp. Biochem. Physiol.
    48A:263-276.
INOUYE, D.W., N.D. FAVRE, J.A. LANUM, D.M. LEVINE, J.B. MYERS,
    M.S. ROBERTS, F.C. TSAO, and Y.Y. WANG. 1980. The effects
    of nonsugar nectar constituents on estimates of nectar
    energy content. Ecology 61:992-996.
JANZEN, D.H. 1977. Why don't ants visit flowers? Biotropica
    9:252.
PERCIVAL, M.S. 1961. Types of nectar in angiosperms. New
    Phytol. 60:235-281.
REHR, S.S., D.H. JANZEN, and P.P. FEENY. 1973. L-dopa in
    legume seeds: A chemical barrier to insect attack. Science
    181:81-82.
RHOADES, D.F., and J.C. BERGDAHL. 1981. Adaptive significance
    of toxic nectar. Amer. Nat. 117:798-803.
RIPER, W. van. 1960. Does a hummingbird find nectar through
    its sense of smell? Sci. Am. 202:157 et ff.
ROBINSON, T. 1968. The Biochemistry of Alkaloids. Berlin:
    Springer-Verlag.
SCOGIN, R. 1979. Nectar constituents in the genus *Fremontia*
    (Sterculiaceae): Sugars, flavonoids and proteins. Bot. Gaz.
    140:29-31.
SCOGIN, R. 1980. Floral pigments and nectar constituents of tw
    bat-pollinated plants: Coloration, nutritional, and energeti
    considerations. Biotropica 12:273-276.
SIMPSON, B.B., J.L. NEFF, and D. SEIGLER. 1977. *Krameria,*
    free fatty acids and oil collecting bees. Nature 267:150-1!
SOKAL, R.R., and F.J. ROHLF. 1969. Biometry. San Francisco:
    W.H. Freeman.
STILES, F.G. 1976. Taste preferences, color preferences, and
    flower choice in hummingbirds. Condor 78:10-26.
VOGEL, S. 1969. Flowers offering fatty oil instead of nectar.
    Abstracts XI Int. Bot. Congress, Seattle, p. 229.
VOGEL, S. 1971. Oelproduzierende Blumen, die durch ölsammelnde
    Bienen bestaubt werden. Naturwissenschaften 58:58.
VOGEL, S. 1974. Ölblumen und ölsammelnde Bienen. Academie
    der Wissenschaften der Literatur Mainz: Tropische und
    Subtropische Pflanzenwelt 7:1-547.
VOGEL, S. 1976. *Lysimachia*: Oelblumen der Holarktis. Natur-
    wissenschaften 63:44-45.
VOSS, R., M. TURNER, R. INOUYE, M. FISHER, and R. CORT. 1980.
    Floral biology of *Markea neurantha* Hemsley (Solanaceae),
    a bat-pollinated epiphyte. Amer. Midl. Nat. 103:262-268.
WALLER, G.D. 1972. Evaluating responses of honey bees to
    sugar solutions using an artificial-flower feeder. Ann.
    Ent. Soc. Amer. 65:857-862.

WATT, W.B., P.C. HOCH, and S.G. MILLS. 1974. Nectar resource use by *Colias* butterflies. Oecologia (Berl.) 14:353-374.

WILLMER, P.G. 1980. The effects of insect visitors on nectar constituents in temperate plants. Oecologia (Berl.) 47:270-277.

WYKES, G.R. 1952a. The preferences of honey bees for solutions of various sugars which occur in nectar. J. Exp. Biol. 29:511-518.

WYKES, G.R. 1952b. The sugar content of nectars. Biochem. J. 53:294-296.

ZIEGLER, H. 1956. Untersuchungen über die Leitung und sekretion der Assimilate. Planta (Berl.) 47:447-500.

COURTSHIP PHEROMONES: EVOLUTION BY NATURAL
AND SEXUAL SELECTION

Stevan J. Arnold

Department of Biology and
Committee on Evolutionary Biology
University of Chicago
Chicago, Illinois

Lynne D. Houck

Department of Education
Field Museum of Natural History
and Department of Biology
University of Chicago
Chicago, Illinois

*Courtship pheromones are defined as chemical signals that
are transmitted between sexual partners during courtship.
Male courtship pheromones are found in a surprisingly diverse
array of animals, but are most common among species of
arthropods and salamanders. Methods of delivering the courtship pheromone to the female vary; there are male butterflies
that dust the pheromone directly on the female's antennae,
scorpionflies and crickets that produce a gustatory offering
which the female ingests during copulation, and salamanders
that apply a glandular secretion to the female's nares or
introduce it directly into her circulatory system before sperm
transfer.*
*Sexual selection is strongly implicated in the evolution of
courtship pheromones since, in most instances, these chemicals
probably affect male mating success. Natural selection may
also affect pheromone evolution to the extent that these
pheromones influence the fertility of females and the mortality
of both sexes. The evolutionary elaboration of courtship
pheromones is affected differently by sexual selection than by
natural selection. The nature of this difference can be
approached by experimental studies that assay the effects of
courtship pheromones on fitness components of both males and
females, and by comparative studies that document the actual
course of evolution.*

> *"...like the birds and beasts which,*
> *attracted by the gaudy appurtenances of sex,*
> *unwittingly perpetuate their species..."*
>
> *Jan Morris, Farewell the Trumpets*

INTRODUCTION

In a diverse group of animals, including insects, pseudoscorpions, fish, mammals and salamanders, the gaudy appur-
tenances of sex include chemicals that do more than simply
attract sexual partners. The chemicals that concern us here
have sometimes been called "aphrodisiacs" but, since sexual
persuasion is only one of many possible effects, we use the
less loaded term "courtship pheromones." These chemicals
appear to be distinct from sex pheromones that act at a distance and merely attract one sex to the other. Common denomin-
ators among the various animal groups employing male courtship
pheromones are: (1) pheromones are produced from sexually
dimorphic, male glands that appear to be used only in courtship, (2) the male delivers the chemical to the female after
their initial encounter but before sperm transfer, and (3)
the pheromone is delivered directly to the female, sometimes
by active application on female chemoreceptors or by injection
into the female's circulatory system. Unlike sex pheromones
or attractants, which have profound economic consequences in
agricultural pest control, courtship pheromones have received
relatively little biochemical scrutiny. In this paper, we brief
review some diverse instances of male courtship pheromone
delivery and present a conceptual framework for organizing
comparative and experimental studies of pheromone evolution.
We also review some recent experimental work with particular
emphasis on our own study organisms, salamanders.

Our principal concern will be with the selection process
that affects the evolution of courtship pheromones rather than
with the immediate physiological effects of chemicals on the
female. The reason for this focus is that we can expect both
extremely rapid evolution of pheromones and delivery systems,
as well as a great diversity of evolutionary outcomes, if
variation in courtship pheromones affects male mating success.
In contrast, if other components of fitness are affected by
pheromones, theoretical results lead us to expect a much less
dramatic evolutionary process (Fisher 1930; O'Donald 1980;
Lande 1980, 1981; Kirkpatrick 1982). Our aim in this paper
is to illustrate how a focus on modes of selection can be
used to organize existing information and how it can inspire
new experimental and comparative studies of courtship pheromones.

SEXUAL VERSUS NATURAL SELECTION

As Ghiselin (1974:130) has remarked, very few modern
authors use the term "sexual selection" as Darwin used it. Most
contemporary authors incorrectly treat sexual selection as a
subcategory of natural selection. Darwin (1859, 1871), however,
repeatedly contrasted the two forms of selection and clearly
viewed them as separate, often opposing processes. For example,
"In regard to structures acquired through ordinary or natural
selection, there is in most cases ... a limit to the amount of
advantageous modification in relation to certain special ends;
but in regard to structures adapted to make one male victorious
over another either in fighting or in charming the female, there
is no definite limit to the amount of advantageous modification;
so long as the proper variations arise, the work of sexual
selection will go on" (Darwin 1871:278). The two agents of
sexual selection, females and rival males, will exert their

effects by influencing the number of mates that bear the progeny of each male, *i.e.*, by affecting male mating success.

Regardless of whether one adheres to Darwin's terminology, it is critical to ask which component of fitness is affected by the pheromone, since this can have a major effect on the evolutionary process. Four major possibilities, outlined below, are that chemical delivery during courtship affects (1) mating success (the number of mates that bear progeny sired by the male), (2) the fertility of mates (the overall number of progeny or rate of production), (3) the survivorship of the mate(s) until the production of progeny, and (4) the survivorship of the pheromone-producing male. Variation in the first component, mating success, constitutes sexual selection while variation in the other aspects, fertility and survivorship, constitutes natural selection (Darwin 1859, 1871; Ghiselin 1974; Lande 1980, 1981; Wade and Arnold 1980). Of course, a courtship pheromone may affect more than one component of fitness.

We first consider how male chemical delivery during court-ship might affect these four components of fitness, and then examine the different evolutionary consequences expected under sexual versus natural selection.

Male mating success. A chemical produced by males during the mating season might have one or more of the following effects on mating success: (1) The chemical might enhance discovery of, or encounter with, females. Although female-produced sex attractants are common, especially among insect species, many examples are known in which males broadcast pheromones that attract females (*e.g.*, Jacobson 1965:39-48; Shorey 1973). Such male chemicals might evolve by sexual selection but would not be courtship pheromones if they merely

attract the female. (2) The male's pheromone might exert its
effect after the initial encounter with the female by promoting
the probability of insemination. Thus the effect may be to
persuade the female to court. Such chemicals are sometimes
called "aphrodisiacs." [Birch (1974) cautions against the
anthropomorphic implications of this term, and we use it here
only in a restricted sense, not implying guaranteed courtship
success.] We later discuss a number of experimental studies
of putative aphrodisiacs in butterflies, mammals and salamanders.
Alternatively, the pheromone may promote insemination by
enhancing coordination of partners during sperm transfer. (3)
A final possibility is that the male's chemical delivery
promotes paternity after the act of insemination. For example,
the male's pheromone could promote sperm transport by inducing
a quiescent state in the female. Paternity also may be enhanced
by so called "anti-aphrodisiacs" that either induce a sexually
nonreceptive state in the female (Barth and Lester 1973; Swailes
1971; Gwadz 1972; Leopold, *et al.*, 1971; Riemann, *et al.*, 1967)
or repel rival males from the inseminated female (Happ 1969;
Gilbert 1976; Hirai, *et al.*, 1978).

Fertility and survivorship of mates. It is possible that
the male chemical will affect mate fertility or survivorship
rather than, or in addition to, mating success. This is most
likely in instances where the female actually ingests the
chemical (gustatory transmission) since a considerable
energetic transfer is conceivable. Thus, in many cockroaches,
crickets and katydids, the female feeds on products from a
specialized male gland and, apparently because she is feeding,
the male is able to copulate with her (Roth and Dateo 1966,
Alexander and Brown 1963, Thornhill 1976a). One effect of
this chemical transfer could be promotion of male mating

success if males that produce gustatory secretions are more
likely to complete insemination (Alexander 1964). Males with
substantial glandular offerings also might increase the
fertility or survivorship of their mates. The semen or
spermatophore itself may transfer materials and energy as
well as gametes to the female (Thornhill 1976a, Gwynne, 1981).
Transfer of nutrients via the spermatophore has been demonstrated
in three species of nymphalid butterflies where male-derived
substances were later found in the developing ova of inseminated
females (Boggs and Gilbert 1979). In many species of caddisflies (Trichoptera), the male transfers a protein-rich secretion
to the female during copulation (Svensson 1972), and the protein
source is used by the female as an energy supply. Similarly,
some scorpionflies offer a salivary secretion to a potential
mate (Thornhill 1976a).

   Survivorship of males. Finally, chemical delivery could
have negative as well as positive effects on individual male
fitness. These negative effects may constitute the opposing
selective forces that eventually halt the evolutionary
elaboration of the delivery system. For example, the system
may actually reduce the survivorship component of male fitness
due to costs of manufacture and maintenance or because predators
as well as mates are attracted. In analogous examples,
parasitoid flies orient to cricket songs (Cade 1975), predatory
bats locate male frogs from their mating calls (Tuttle and
Ryan 1981) and predatory female fireflies attract and consume
the males of other species by mimicking the flash responses
of the prey's own females (Lloyd 1975). These examples suggest
that predators could cue on sex attractants or courtship
pheromones. Furthermore, the chemical may attract rival males
as well as potential mates. This apparently is a common

liability associated with pheromone broadcasting. In many
insects, males as well as females are attracted by pheromones
produced by conspecific males (see examples in Jacobson 1965).

EVOLUTIONARY CONSEQUENCES

A pheromone that exclusively affects mating success is
analogous to frog vocalization or to the peacock's fabulous
display. Such a pheromone will have very different evolutionary
consequences than one that affects mate fertility or survivorship. As Darwin (1859, 1871) pointed out, such male attributes
may evolve under the force of a purely sexual advantage.
Theoretical models of sexual selection show that the male's
tail or his pheromone can be elaborated during evolution even
though the female receives no direct or immediate advantage
from mate choice (Lande 1980, 1981; O'Donald 1977, 1980;
Fisher 1930; Kirkpatrick 1982). As Fisher (1930) first showed,
there can be an accelerating coevolution of a male trait and
the female response to this trait. Using pheromones as an
example, both the male pheromone and the female reaction to
this pheromone can increase during evolution since, in each
generation, the males with the most effective pheromone are
more likely to inseminate both the most resistant and the
most responsive females. If there is heritable variation in
male pheromones and in female response to the pheromone, the
mate selection process can create and maintain a genetic
correlation between the two attributes (Lande 1981, Kirkpatrick
1982). The evolution of female responsiveness thus can occur
purely as a genetic corollary of sexual selection on male
pheromone, not because the females benefit or suffer in any
direct way from the pheromone.

Lande (1980, 1981) and Kirkpatrick (1982) have shown that
the evolutionary outcome of this sexual selection process is

indeterminate. The equilibrium values of average male
pheromone production and average female resistance will vary
according to the initial conditions in the population: there
is no unique evolutionary outcome. Consequently, in any
radiation in which pheromonal evolution by sexual selection
is a recurrent happening, we can expect a tremendous variety
of evolutionary outcomes. Furthermore, the variety will be
largely historical and nonadaptive.

The evolutionary expectation is quite different if the
male's pheromone affects the fertility or survivorship of his
mates (Lande 1981). In this case there will be a unique
equilibrium of male pheromone production and average female
reaction. Differences between species and among higher taxa
may be less extensive than differences produced purely by
sexual selection.

SURVEY OF MALE CHEMICAL DELIVERY DURING COURTSHIP

In the following sections we consider a variety of
examples in which males deliver chemicals during courtship.
Male courtship pheromones are found in a diverse group of
animal species, but are especially common in many orders of
insects, including species of scorpionflies (Thornhill 1976a,
1976b), caddisflies (Svensson 1972), butterflies and moths,
cockroaches and crickets, water bugs, bees, flies, and
lacewings (see reviews by Jacobson 1965, Birch 1974, and
Shorey 1973). Male chemical delivery during courtship also
is known for pseudoscorpions, some fish and mammals, and
for most species of salamanders. It is not our goal to
present an exhaustive survey of all known examples of chemical
persuasion during courtship. Instead, we focus primarily on
studies that experimentally test for effects of male pheromones

on different components of fitness. Such experimental
studies can indicate whether the pheromone delivery system
is affected purely by sexual selection or whether natural
selection on males and females also affects pheromonal evolution.
Unfortunately, no species has been completely analyzed from this
point of view. Nevertheless, the available studies indicate
how one might address the problem of partitioning the effects
of selection into the categories of natural and sexual selection.

Insects. Pheromonal communication among insect species is
widespread and diverse. Even restricting our consideration
to male courtship pheromones (those produced only after the
male and his potential mate have been brought together), there
still are numerous examples from a variety of insect species
(see Jacobson 1965, Birch 1974, and Shorey 1973). Some of the
most thorough work that demonstrates the nature and effects of
male courtship pheromones was conducted with the day-flying
Queen butterfly, *Danaus gilippus* (Brower and Jones 1965; Brower,
et al., 1965; Pliske and Eisner 1969). During the courtship
season, the male is visually attracted to the female as she
flutters through the air (males are not discriminatory and
also may approach other males or even falling leaves; Myers
1972). The male hovers above the female and extrudes a pair
of specialized glands called "hairpencils" (fig. 1). When
the secretions from the hairpencils are delivered in the
vicinity of the female's antennae, she tends to alight on the
ground or on nearby vegetation. The male follows the female
and continues to release secretions until he achieves a
copulatory position. In an experimental study of the Queen
butterfly conducted in a natural habitat, Myers and Brower
(1969) showed that interference with male chemical delivery
(either by removing the male's hairpencils or by blocking the

sc

h

d

female's antennal receptors) resulted in substantially reduced courtship success. Thus the pheromone released from the hairpencils affects the mating success component of male fitness, so it is extremely likely that this courtship pheromone evolved by sexual selection. Male *D. gilippus* have other specialized glands, wing pockets, which also may affect male courtship success; the function of these glands, however, has not been conclusively demonstrated (Myers 1972).

In another example of chemical use during courtship, Tinbergen *et al.* (1943) showed that a male satyriid moth, *Eumensis semele,* actually moves the female's antennae between his forewings so that her antennae directly contact the specialized scales that constitute his scent patches. This direct application of courtship pheromones is exceptional since most lepidopterans apparently rely on airborne chemical delivery.

Some insect aphrodisiacs are gustatory in that chemical secretions produced by the male are ingested by the female. While the female feeds on the secretions, she usually assumes

---

FIGURE 1. *The male Queen butterfly (Danaus gilippus)
releases courtship pheromone from glandular
hairpencils at the tip of his abdomen. (A)
The male hovers and bobs in front of the
female (stippled) and his fully everted hairpencils dust pheromone (a ketone) directly
on the female's antennae. (B) Diagrammatic
longitudinal section through the glandular
base of the male's retracted hairpencil. The
individual hairs (h) of the hairpencil are
hollow, perforated processes from trichogen
secretory cells (sc.) Globular secretions
called "dust" (d) accumulate on the hairs.
During courtship the glandular base everts
and the hairs splay out. A retractor muscle
(not shown) attaches to the proximal end of
the bundle (on the left). (After Brower et al.,
1965).*

a copulatory position and the male may inseminate her at this time. Examples of gustatory aphrodisiacs are reviewed by Birch (1974), and are found in at least one species of tephritid fly, 25 species of crickets and cockroaches, and probably in many species of coleopterans. Although some gustatory pheromones are known to promote male mating success, so that sexual selection is implicated in their evolution, they may have positive effects on other aspects of fitness as well. When there is extensive feeding on a male gland, as in many crickets, the female's clutch size might be increased.

Pseudoscorpions. In cheliferid pseudoscorpions, females appear to orient to pheromones emanating from the male's "ram's-horn" organs (Weygoldt 1966, 1969). These organs are evaginated just before sperm transfer (fig. 2). The ram's-horn organs are not glandular, but they probably are coated with pheromones from glands that empty into the genital chamber. The female contacts these organs with her pedipalps prior to sperm transfer, and this is probably when chemo-reception takes place.

Vertebrates. Although pheromonal communication between the male and female is common during vertebrate reproduction, most studies focus on the nature and effect of chemical signals produced by the estrous female. Instances of chemical communication directed from the male to the female are less widely studied and so the extent of precopulatory, male-to-female pheromonal communication is not known. Except for most species of salamanders, there are relatively few examples among vertebrates, especially for male chemical delivery that occurs only during courtship. Male courtship pheromones are known for some species of fish and mammals, however, and further studies may show the phenomenon to be more widespread than previously realized.

In glandulocaudine fish of the family Characidae, males of different species show a variety of glandular pockets and scales at the base of the tail (Nelson 1964). The behavior of the male during courtship suggests that the male wafts pheromones from these caudal glands to the female (fig. 3). Compared with other characid fish, the glandulocaudines have undergone an explosive radiation in courtship glands and other sexually dimorphic structures (Eigenmann and Myers 1929), a radiation undoubtedly promoted by sexual selection.

Among some species of mammals, odors from sexually mature males are known to affect female behavior during courtship. In the domestic pig, an estrous female responds to pressure on her dorsum by exhibiting a copulatory stance called the "standing reaction." This reaction terminates the pre-copulatory phases of sexual behavior since a male can attempt intromission with a female that assumes a "standing" (braced and motionless) posture. Signoret (1976) reports that, among estrous sows, the standing reaction can be induced in 48% of females that are touched on the back by a human experimenter (no boar is present). In comparison, this response can be induced in 100% of estrous females if there is a boar present. This increase is largely attributable to male odor since the standing reaction occurs in 90% of estrous sows even when they cannot see or touch the boar. Of 90 estrous sows that originally failed to exhibit a standing reaction in response to a human experimenter, 56 (62%) of these females gave a positive standing reaction when placed in a boar's pen even though the boar was absent (Signoret and du Mesnil du Buisson 1961). Odors from fluid collected from the preputial pouch of the male were just as effective as total male odor in eliciting the female standing response (*Ibid*). Because effects on male mating success have been experimentally verified, this is the best mammalian example

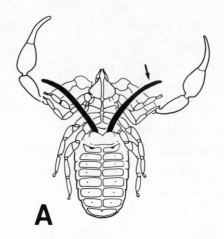

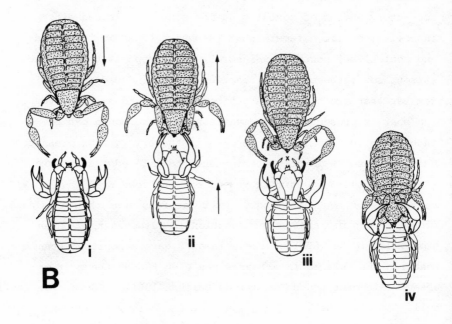

of a sexually selected courtship pheromone.

The effects of chemical delivery by other mammalian males have not been as clearly demonstrated as the effects of boar odor on estrous sows. One conspicuous example of chemical delivery is urine spraying, a behavior that is relatively common among lagomorphs and among some histricomorph rodents (porcupines, guinea pigs, agoutis and their allies). Odors in general are significant among these social species, and groups of individuals commonly share odors by marking each other and themselves. In a review of studies on the European wild rabbit, *Oryctolagus cuniculus,* Bell (1980) reports that, in 15 cases where urine spraying actually was observed in the

FIGURE 2. *Delivery of a putative courtship pheromone during courtship and sperm transfer in the pseudoscopion* Dactylochelifer latreillei. *(A) Ventral view of a male showing his everted ram's-horn organs (black). These organs are lateral diverticulae of the male's genital atrium. The ram's horn organs are not glandular and consist of a thin epithelium bounded by a cuticle. They apparently are coated with a courtship pheromone produced by one of the many glands emptying into the genital atrium. (After Weygoldt 1969). (B) The sperm transfer process: (i) The female (stippled) approaches the male as he displays his ram's-horn organs (black) and vibrates his entire body. The tips of her pedipalps (large, pincher-like appendages) are studded with setae that are thought to be chemosensory. (ii) The male stops vibrating and the pair moves backward. This activity alternates with the vibratory male display and female approach shown in (i). (iii) The male has deposited his spermatophore (x) on the substrate and has stepped back from it. The female is approaching the male and his spermatophore, apparently orienting to his ram's-horn organs. (iv) The male grasps the female's pedipalps as she settles on top of the spermatophore. (After Weygoldt 1966).*

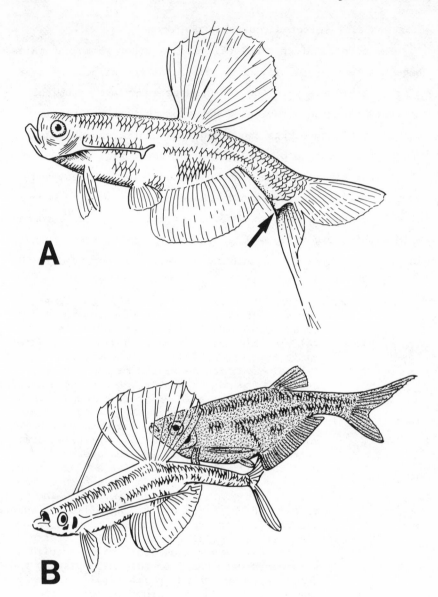

field, the spraying was performed by sexually active, dominant males and was directed toward conspecifics. In only two of the 15 cases, however, did urine spraying occur during courtship. Kleiman (1974) suggests that urine spraying during histricomorph courtship may be a method of reducing agonistic behavior in the female by clearly identifying the male as a familiar member of her social group. Although the male's use of urine may have a positive influence on female receptivity, other contexts of urine delivery preclude its consideration strictly as a courtship pheromone.

The evolution of male odors that result in female behavior conducive to copulation can be explained in terms of sexual selection since these odors apparently promote the mating success of the individual male. Although Signoret (1976) reports experimental demonstration of the effects of male odors in the domestic pig, a direct field test to evaluate the effect of odor production on male mating success would be desirable. Evaluations in field situations unfortunately can be difficult since they require quantification of natural variation in odor produced among males and since other factors, such as natural variability in female receptivity, might be difficult to measure. Although field tests are desirable,

FIGURE 3. *Male* Corynopoma riisei, *like many other glandulocaudine fish, appear to release a courtship pheromone from their caudal glands. (A) The male has highly modified fins and the extraordinary operculum is shaped like a paddle. The site of the glandular pouch that constitutes the caudal gland is indicated with an arrow. Lateral movements of the tail apparently flush pheromone out of the pouch and waft it toward the female. (B) During courtship the male swims in a figure-of-eight pattern in front of the female (stippled), a movement that probably wafts courtship pheromone to her. (After Nelson 1964).*

laboratory tests of odor-related mating success in males also
can be informative. Male odor could be artificially enhanced
or eliminated, for example, or the receptivity of anosmic
females could be compared with that of intact females. In
the following section, we discuss experimental data of this
sort for salamanders.

CHEMICAL DELIVERY DURING SALAMANDER COURTSHIP

Most salamanders are nocturnal and, not surprisingly, many
aspects of their sexual behavior are chemically mediated. The
functions of chemical signals include mate attraction,
orientation and coordination during sperm transfer, and sexual
persuasion. Twitty (1955, 1961), for example, found that
male newts of the genus *Taricha* can identify conspecific
females even by their skin secretions alone. Male *Taricha*
are attracted to sponges soaked with secretions from conspecific
females, but not to sponges soaked with secretions from hetero-
specific females. As in most salamanders, *Taricha* males nose
the female before initiating courtship, and species and sex
identification probably is accomplished at this early stage.
It is not known whether the female has evolved a special sex
attractant pheromone or whether the male simply responds to
normal, epidermal secretions.

Chemical signals from the male play an important role
during the final stage of courtship, when coordination of
male-female actions is crucial during the intricate process
of sperm transfer. In most salamanders, sperm is transferred
via spermatophores attached to the substrate. Commonly, the
female finds the spermatophore only by following the male and,
among aquatic species, by orienting to pheromones released by
glands in the male's cloaca. During sperm transfer in the

aquatic-breeding tiger salamander (*Ambystoma trigrinum*),
for example, the male walks with his tail raised at a 45°
angle and the female orients to his cloacal glands (Arnold
1976)(fig. 4). Sexual selection probably contributed to the
evolution of these glands, but experimental tests for an
effect on mating success have not been performed.

There is evidence in salamanders that sexual persuasion
of the female is accomplished by male glands that are employed
during the intermediate stage of courtship. The length of this
stage varies within and among species but is defined as the time
after individual identification but before attempts at sperm
transfer. During this intermediate stage, males deliver
secretions from specialized epidermal glands that are present
only in adult males and are most highly developed during the
courtship season. Commonly, these courtship glands are located
on the male's head and are rubbed directly on the female's nares.
In aquatic species, secretions from cloacal glands may be wafted
towards the female's nose by the male's tail fanning behavior.
Although courtship glands are extremely common and diverse in
salamanders (Noble 1931, Arnold 1977), their persuasive effects
have been experimentally studied only in two genera of newts
(*Notophthalmus* and *Triturus*) and in the plethodontid genus
Desmognathus. We review these experimental data below.

In the eastern North American newt, *Notophthalmus viridescens,*
the male adjusts his courtship to the responsiveness of the
female (Humphries 1955; Arnold 1972, 1977; Verrell, pers.
comm.). The male's courtship display is extremely brief if the
female indicates receptivity by actively approaching the male,
or if she does not swim away when the male approaches her. In
these situations the male gives a short display with undulating
tail, and, if the female responds, he proceeds directly with
spermatophore deposition. In contrast, if the female is not

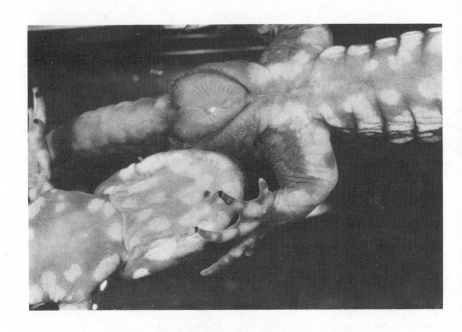

immediately receptive and she swims away from the approaching
male, he pursues her and attempts to capture her with his
hind limbs. If he succeeds, he holds the female in amplexus
and for an hour or more delivers secretions from courtship
glands before attempting sperm transfer (Arnold 1977, fig. 5).
Two kinds of courtship glands are employed during amplectic
courtship. A sexually active *N. viridescens* male has genial
glands on his cheeks (fig. 5). The development of these
glands is hormonally regulated (Pool and Dent 1977), and their
product is a sulfated mucin that is released in response to
cholinergic stimulation of myoepithelial cells (Pool *et al.*,
1977). Secretions from the genial glands are repeatedly
rubbed over the female's nares. In addition, glands in the
male's cloaca are everted during courtship. Pheromones from
these cloacal glands probably are wafted to the female's nares
by fanning movements of the male's recurved tail. Dye placed
in the water near the male's cloaca is transported by water
currents produced by the male's tail movements and the dye

FIGURE 4.   *The female tiger salamander (*Ambystoma tigrinum*)
appears to orient to the male's cloacal glands
during sperm transfer. A pair of courting
salamanders was photographed from below, through
the floor of a glass-bottomed aquarium. ABOVE.
The female nudges the male's cloaca as he deposits
a spermatophore. The whitish sperm mass, which
is perched on the top of a transparent pyramidal
base, is visible at the center of the male's
cloaca. The papillae of his cloacal glands
fringe the posterior edge of his cloaca. BELOW.
The male moves forward away from his spermatophore.
Only the white sperm mass at the apex of the
clear spermatophore is distinctly visible. The
female follows the male, apparently orienting to
a courtship pheromone released from his cloaca.
She will settle on top of the spermatophore
and remove the sperm mass with her cloacal lips.*

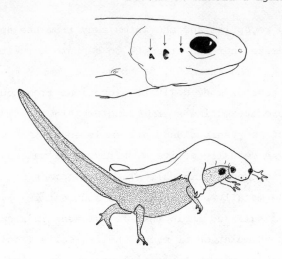

FIGURE 5.   The male red-spotted newt (Notophthalmus
            viridescens) produces a courtship pheromone
            in glands on his cheeks. ABOVE. The male's
            head showing the apertures of his genial
            glands. BELOW. During courtship the male
            clasps the female (stippled) and repeatedly
            rubs his glands across her nares. (From
            Arnold 1977). Reprinted with permission, from
            "The Reproductive Biology of Amphibians", 1977,
            Plenum Publishing Corporation, New York.

gathers about the female's snout (Verrell, pers. comm.).
Throughout the lengthy amplexus, direct delivery of genial
gland secretions alternates with wafting delivery of cloacal
secretions. The complex temporal structure of these and other
activities during the amplectic courtship is described by
Zeller (1890), Jordan (1891), Humphries (1955) and Arnold
(1972).

    Experimental and observational studies of Notophthalmus
viridescens indicate that chemical delivery during amplectic cour
ship promotes sperm transfer and hence male mating success.   Rogc
(1927) found that the female would not follow the male for
sperm transfer after amplexus if the male's genial glands
were artificially occluded or if the female's nares were

plugged. Verrell (pers. comm.) found that sperm transfer was positively correlated with two indices of genial gland application.

Experimental studies of European newts of the genus *Triturus* also suggest that chemical delivery during courtship enhances mating success. Unlike *Notophthalmus, Triturus* never show amplexus behavior and thus the male never restrains the female during courtship. Other aspects of courtship behavior are similar between these two genera, however, and *Triturus* may be derived from a *Notophthalmus*-like ancestor (Salthe 1967, Arnold 1972, Halliday 1977). During courtship, a male *Triturus* performs an elaborate tail fanning display in front of the female as a precursor to sperm transfer. As in *Notophthalmus,* the male's cloacal glands are everted during this display and tail fanning creates a water current that probably wafts cloacal pheromones to the female's nose (Tinbergen and Ter Pelkwijk 1938, Halliday 1975b). In the most studied species, *T. vulgaris,* the male may deposit as many as seven spermatophores when courting a female, although the average per courtship is two (Halliday 1977). After each spermatophore deposition and transfer attempt, the male reverts to his tail fanning display. Remarkably, the continuation of male courtship does not depend on whether the female picks up sperm from spermatophores. The male is just as likely to proceed with courtship after a successful transfer as he is after an unsuccessful attempt (Halliday 1975a). The male does change the timing and content of his displays in relation to the number of spermatophores already deposited. Apparently the male can manufacture only a limited number of spermatophores each day and, as the male approaches the end of his spermatophore supply, his display is slower and more elaborate (Halliday 1976). By manipulating anaesthetized females so that their "behavior" indicated

receptivity, Halliday (1975a) was able to show that temporal
changes in the male's courtship displays are due to the male's
state as well as the female's behavior. As the male exhausts
his supply of spermatophores, he requires more tactile cues
from the female before he will deposit another spermatophore.
The depletion of the male's spermatophore supply apparently
accounts for the change in courtship choreography.

The tail fanning aspect of the male *Triturus* display may
have a cumulative chemical effect on the female. Females tend
to walk past the spermatophores deposited early in courtship,
and spermatophores that are missed invariably are reproductive
failures. Females are more likely to retrieve sperm from
spermatophores deposited late in courtship, after the male
has delivered much pheromone to the female. Thus female
T. vulgaris retrieved sperm from only 29% of the first
spermatophores deposited during courtship, but they retrieved
sperm from 62% of the third spermatophores (Halliday 1977).
This suggests that chemical delivery during the tail fanning
display may have a cumulative, aphrodisiac-like effect on the
female. Alternatively, Halliday (1977) proposed that the
tendency of females to miss early spermatophores represents a
form of mate assessment. By missing the first spermatophores
deposited by a male, females might avoid insemination by males
with relatively poor spermatophore supplies.

In contrast to the aquatic courtship typical of salamandrids
and ambystomatids, most plethodontid (lungless) salamanders
court on land. Sexually mature males of most plethodontid
species possess a submandibular or mental (*mentum* = chin)
gland (fig. 6) that is maximally developed during the courtship
season (see Lanza 1959 and Sever 1976a). Glandular development
is hormonally controlled; Sever (1976b) showed that, in *Eurycea
quadridigitata,* even females would develop a mental gland when

injected with testosterone. Sexually active males of many
plethodontid species also have specialized teeth that differ
from the female's teeth both in size and shape (fig. 7). The
male's mental gland is the primary source of courtship phermones
and, in species where males also have specialized teeth, these
teeth are an integral part of the chemical delivery system.

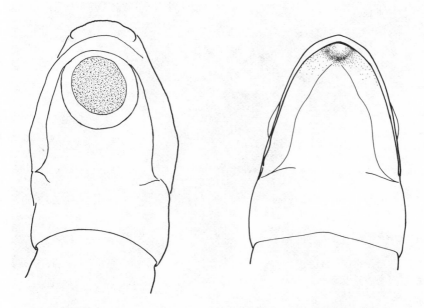

FIGURE 6.   *Male plethodontid salamanders have courtship*
glands on their chins. LEFT. Plethodon
jordani *males have large, disc-shaped mental*
glands. Such large glands are characteristic
of species that deliver the putative courtship
pheromone by slapping the gland on the female's
snout, so that she inhales the secretion. (From
Arnold 1972). RIGHT. Male Desmognathus fuscus
have small mental glands, set on a platform at
the tip of the chin. Small glands are found
in species that introduce the putative court-
ship pheromone directly into the female's
circulatory system by abrading her skin with
protruding premaxillary teeth. (After Noble 1927).

There are four modes of chemical delivery during plethodontid courtship. The first two modes are termed "pulling"
and "snapping." "Pulling" consists of a series of rapid
strokes with the male's chin pressed down on the female's body

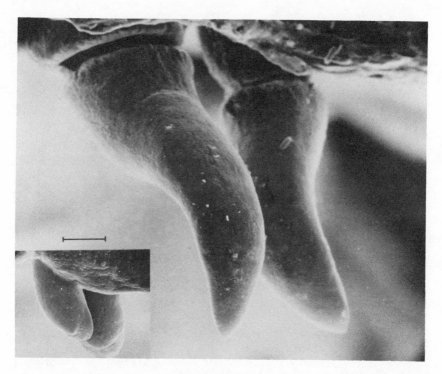

FIGURE 7.   Scanning electron micrograph of the premaxillary
            teeth of Desmognathus ochrophaeus. The adult
            male has monocuspid premaxillary teeth that are
            roughly three times the size of the female's
            bicuspid teeth (inset). The male's teeth are
            used to introduce a courtship pheromone into
            the female's circulatory system and also in
            biting combat with other males. The male's
            and female's teeth are shown at the same scale
            (bar = 0.5 mm).

while "snapping" consists of a single very forceful stroke,
accomplished by a sudden snapping action of the male's body
which may fling the male away from the female (fig. 8).
Pulling and snapping often are used alternately during courtship and occur in species with relatively small mental glands
and protruding premaxillary teeth. Both actions apparently

FIGURE 8.   Delivery of courtship pheromone in Desmognathus
            ochrophaeus. The male presses his protruding
            premaxillary teeth and mental gland against
            the female's flank as she straddles his
            undulating tail. From this position the male
            forcefully snaps his head downward, abrading
            the female's skin at the site rubbed with
            secretion from his mental gland.

have the effect of "vaccinating" the female with mental gland
secretions (Organ 1961, Arnold 1972). These two actions occur
in many diverse genera and tribes of plethodontids, and
probably represent the ancestral delivery modes (Arnold 1977).
The third delivery mode, "slapping," occurs only in species
that have relatively large mental glands but lack protruding
premaxillary teeth. These species slap the mental gland on
the female's snout and the female apparently inhales the
secretions through her nares (Arnold 1976). The fourth
delivery mode, biting, is known to occur only in one species,
Desmognathus wrighti (Houck 1980). The male bites the female
as the initial courtship action and holds on tenaciously. The
males possess highly modified mandibular teeth, and the
secretions from the mental gland apparently are introduced
through perforations in the female's skin produced by these
teeth.

The only experimental analysis of the effect of male
chemical delivery during plethodontid courtship was conducted
with the salamander *Desmognathus ochrophaeus*. Details of
the experiment are reported elsewhere (Houck and Arnold, in
preparation), but are summarized here. Since courtship and
sperm transfer in *D. ochrophaeus* occur on land, aspects of
chemical delivery therefore are somewhat different from those
reported for the aquatic breeding newts. During the courtship season, an adult male *D. ochrophaeus* has specialized
monocuspid premaxillary teeth that are about three times
longer than the female's bicuspid premaxillary teeth (fig.
7; also see Noble 1931). Males of this species also have a
mental gland, and secretions from this gland are applied to
the female primarily during the two male behaviors termed
pulling and snapping (described above). The experiment
investigating chemical persuasion during *D. ochrophaeus* court-

ship was designed to test whether a male's ability to in-seminate a female is affected by (1) the lack of a mental gland, (2) the lack of premaxillary teeth, or (3) lack of both the mental gland and the premaxillary teeth. The experiment involved surgical treatment of three groups of males, each with its own control group of sham-operated males (mental gland removed, premaxillary teeth removed, both mental gland and premaxillary teeth removed), as well as a control group of intact males. Surprisingly, none of the surgically treated groups of males showed any impairment in insemination success.

The lack of demonstrable effects of mental gland secretions is surprising since the courtship pattern in *D. ochrophaeus* fits our initial conditions for a situation where chemical persuasion is suspected: (1) the males possess specialized glands and associated structures that are well-developed only during the courtship season, (2) males deliver glandular secretions only during courtship, after the initial contact with the female, and (3) the male secretions are applied directly to the female. The experimental results, however, certainly suggest that mental gland delivery does not enhance insemination success. It is nevertheless possible that the gland and sexually dimorphic teeth evolved by sexual selection. For example, it is conceivable that the mental gland secretions promote sperm uptake from the spermatophore after insemination. This could be accomplished very simply by inducing a quiescent state in the female which promotes sperm transfer from the spermatophore to the female's spermatheca, and consequently paternity. Of course this is mere speculation and new experiments will be necessary to determine whether sexual or natural selection is the most plausible agent for the evolution of plethodontid mental glands.

PHEROMONAL EVOLUTION

How do aphrodisiacs and other pheromones evolve? In order
to answer this question we must know more than the physiological
"function" of the pheromone, although this is certainly an
important consideration. To answer even the simpler question of
whether the pheromone evolved by sexual or by natural selection,
it is desirable to marshal evidence from a variety of sources.

Evidence from sexual dimorphism. Darwin (1871) outlined
three criteria that implicate evolution by sexual selection
rather than by natural selection. Sexual selection is
implicated if the character is (a) present only or most highly
developed in males; (b) most highly developed at sexual
maturity and (c) present in highest development during the
breeding season. Even if a character meets all these criteria
there still is some ambiguity about the form of sexual selection.
For example, the antlers of the stag may have evolved due to
advantages conferred in male combat or because females chose
the best endowed stags as mates.

Evidence from behavioral context. Strong hints about the
nature of selection can be derived from the behavioral contexts
in which pheromones and other structures are used. A male
pheromone that is applied directly to the female's olfactory
receptors during courtship is likely to be sexually selected,
either because it promotes the male's mating success directly,
or indirectly if it lessens the chance that the female will
mate with other males. A chemical that is applied to the
female's body, rather than on her receptors, may simply repel
other males and have no effect on the female herself (*e.g.*,
Gilbert 1976). Natural selection also is implicated if there

is a considerable transfer of materials or energy, since effects on fertility or female survivorship are conceivable. Of course, these kinds of evidence are circumstantial and merely suggest possible effects on fitness. Furthermore, the situation can be more complicated if a pheromone or gland affects more than one component of fitness.

Comparative evidence. Only comparative studies can document the actual occurrence of evolution. The best evidence of this kind will come from studies of pheromones in groups with well-established phylogenies. Phylogenetic correlation with social system can give clues about evolutionary history. If, for example, male pheromones tend to occur in species with polygynous mating systems but not in related, monogamous species, we have a *prima facie* case for chemical evolution by sexual selection. Likewise, spectacular phylogenetic diversity may indicate rapid evolution by sexual selection. This expectation comes from theoretical predictions of rapid evolution under sexual selection (Fisher 1930) and of numerous alternative outcomes at equilibrium (Lande 1981). Thus the spectacular plumage diversity in male birds of paradise, which perform elaborate displays at traditional mating grounds and form no pair bond with the female, is almost certainly a product of rapid evolution driven by sexual selection. The rapidity and superficial nature of the divergence among these birds is indicated by the fact that even populations with striking differences in plumage are known to hybridize (Gilliard 1969). We know of no comparable studies of courtship pheromone diversity.

Experimental evidence, genetic analysis and the measurement of selection. Variation, inheritance and selection are necessary,

although not sufficient, conditions for pheromonal evolution;
they are the key issues for analysis of the evolutionary
process. There is much to be learned about each aspect.
Within populations, the study of individual variation both in
pheromone production and in the reaction to pheromones is
pivotal since analyses of inheritance and selection are
predicated on a characterization of such variation. Despite
recent discoveries of optical isomers (enantiomers) and multi-
component pheromones in insects (see review in Ritter 1979),
for example, it still is not clear whether there are stable
individual differences in kinds of pheromones within conspecific populations. Technical limitations may force a
typological attitude: it is easy to assume that all individuals
are identical if individuals can not be assayed. In order to
characterize the structure of sex pheromones in the cockroach
Periplantea americana, Persoons and Ritter (1979) had to
extract intestines from 32,000 insects. When this many
individuals must be pooled for chemical analysis there is
little margin to detect individual differences in pheromones.
We may be missing the raw materials upon which selection acts.

Genetic analysis also confronts formidable obstacles. In
addition to the technical problems of extraction, there are
problems in characterizing individual variation if pheromones
change with age, diet or social situation (*e.g.,* see Bell 1976).
Nevertheless, there are some encouraging developments. Host
races of the larch bud worm, *Zeiraphera diniana*, differ in sex
attractants released by females and the males show corresponding
differences in electrophysiological response by antennal
receptors. F_1 hybrid male antennal responses include both
the responses characteristic of the two parental host races
and a receptor cell that shows maximal response to both
parental pheromones (Priesner 1979). The potential for

screening large numbers of individuals in order to detect
heritable variation in chemoreception is illustrated for
garter snakes (*Thamnophis*) by Arnold (1980, 1981). Furthermore,
there is evidence for coevolution of sex pheromones and
responses to sex pheromones in the pine beetle, *Ips pini*
(Plummer *et al.* 1976). This species shows geographic variation
both in enantiomers that compose the pheromone and in responses
to these enantiomers, making this system an outstanding candidate
for genetic analysis. Because so little is known about genetic
variation both in sex pheromones and the responses they elicit,
we are a long way from a direct test of genetic predictions
made by sexual selection theory. One prediction is that
evolution of courtship pheromones by sexual selection should
create gametic (or linkage) disequilibrium between the pheromones
of males and the response by females (Lande 1981, O'Donald 1980).
This genetic prediction could be tested by assaying for genetic
correlations in natural populations (*e.g.*, by testing for a
statistical regression of daughter's response to pheromone on
father's pheromone characteristic). Falconer (1980) and
Bulmer (1980) discuss estimation techniques.

Experimental tests for the presence of selection on the
pheromone can be arranged by ablating or occluding the glands
that produce the pheromone. In order to test for sexual
selection, the mating success of ablated males can be compared
with an intact group. Such tests can even be performed in the
field (see Brower *et al.* 1965 and Myers and Brower 1969).
Assays of this kind measure the force or gradient of selection
along an artificial character scale ranging from no pheromone
to the mean pheromone characteristic of the population. In
order to measure the actual force of selection acting on the
spectrum of pheromone variation in nature, it would be necessary
to measure both the pheromone characteristics and the mating

success of each male in a sample (Arnold and Wade, in prep.;
Lande and Arnold, in prep.). With appropriate longitudinal
data (individually marked males followed through time), it
should be possible to determine whether pheromones are favored
by sexual selection (*e.g.*, by mate attraction or persuasion)
but disfavored by natural selection (*e.g.*, by attracting
predators). The opposing forces of natural and sexual
selection could act to stabilize the size of the pheromone-
producing gland. Tests and evaluation of selective pressures
ultimately depend on having some means of characterizing the
pheromone chemistry of individuals. If this can be
accomplished, then the relative success of particular pheromone
phenotypes in attracting mates could be assayed. Among insect
species, there are standard insect baiting assays commonly used
in economic applications. Data from assays such as these could
then be used to estimate the actual force of selection exerted
by mate attraction on the male population.

ACKNOWLEDGEMENTS

This work was supported by NSF Grant BNS 80-14151, by
PHS Grant 1 KO4 HD00392-01 and by the Spencer Foundation.
We thank our reviewers for their constructive comments on
the manuscript and Meredyth Friedman for technical assistance.
We are also grateful to Paul Verrell for allowing us to refer
to his unpublished work on *Notophthalmus*.

LITERATURE CITED

ALEXANDER, R.D. 1964. The evolution of mating behavior in
    arthropods. Roy. Entomol. Soc. (London), Symposium
    2:78-94.

ALEXANDER, R.D., and W.L. BROWN, JR. 1963. Mating behavior and the origin of insect wings. Misc. Pub. Mus. Zool., Univ. Michigan 133:1-62.

ARNOLD, S.J. 1972. The evolution of courtship behavior in salamanders. Unpublished Ph.D. Dissertation, Univ. of Michigan. 570 pp.

ARNOLD, S.J. 1976. Sexual behavior, sexual interference and sexual defense in the salamanders *Ambystoma maculatum,* *Ambystoma tigrinum* and *Plethodon jordani.* Z. Tierpsychol. 42:247-300.

ARNOLD, S.J. 1977. The evolution of courtship behavior in New World salamanders with some comments on Old World salamandrids. Pp. 141-183. *In:* The Reproductive Biology of Amphibians, D.H. Taylor and S.I. Guttman, eds. New York: Plenum Press.

ARNOLD, S.J. 1980. The microevolution of feeding behavior. Pp. 409-453. *In:* Foraging Behavior: Ecological, Ethological and Psychological Perspectives, A. Kamil and T. Sargent, eds. New York: Garland Press.

ARNOLD, S.J. 1981. Behavioral variation in natural populations. I. Phenotypic, genetic and environmental correlations between chemoreceptive responses to prey in the garter snake, *Thamnophis elegans.* Evolution 35:489-509.

BARTH, R.H., and L.J. LESTER. 1973. Neuro-hormonal control of sexual behavior in insects. Ann. Rev. Ent. 18:445-472.

BELL, D.J. 1976. Pheromonal communication in rabbits. Paper presented at Second ECRO Congress, Reading, UK. Sept. 1976. Chemorecep. Abs. 5:No. 1.

BELL, D.J. 1980. Social olfaction in lagomorphs. *In:* Olfaction in Mammals. Symp. Zool. Soc. London 45:141-164.

BIRCH, M.C. 1974. Aphrodisiac pheromones in insects. Pp. 115-134. *In:* Pheromones, M.C. Birch, ed. North-Holland Res. Monogr., Frontiers of Biology, Vol. 32. New York: Amer. Elsevier Publ. Co., Inc. 495 pp.

BOGGS, C.L., and L.E. GILBERT. 1979. Male contribution to egg production in butterflies: evidence for transfer of nutrients at mating. Science 206:83-84.

BROWER, L.P., J.V.Z. BROWER, and F.P. CRANSTON. 1965. Courtship behavior of the queen butterfly, *Danaus gilippus berenice.* Zoologica 50:1-40.

BROWER, L.P., and M.A. JONES. 1965. Precourtship interaction of wing and abdominal sex glands in male *Danaus* butterflies. Proc. Roy. Entomol. Soc. London Ser. A 40:147-151.

BULMER, M.G. 1980. The Mathematical Theory of Quantitative Genetics. Oxford: Clarendon Press. 255 pp.

CADE, W. 1975. Acoustically orienting parasitoids: fly phonotaxis to cricket song. Science 190:1312-1313.

DARWIN, C. 1859. The Origin of Species by Means of Natural
    Selection or the Preservation of Favoured Races in the
    Struggle for Life. London: John Murray. 490 pp.
DARWIN, C. 1871. The Descent of Man, and Selection in
    Relation to Sex. London: John Murray. 475 pp.
EIGENMANN, C.H., and G.S. MYERS. 1929. The American Characidae
    Part V. Mem. Mus. Comp. Zool. Harvard XLIII(5):429-558.
FALCONER, D.S. 1980. Introduction to Quantitative Genetics.
    New York: Ronald Press. 365 pp.
FISHER, R.A. 1930. The Genetical Theory of Natural Selection.
    Oxford: Clarendon Press. 272 pp.
GHISELIN, M.T. 1974. The Economy of Nature and the Evolution
    of Sex. Berkeley, Calif.:Univ. Calif. Press. 346 pp.
GILBERT, L.E. 1976. Postmating female odor in *Heliconius*
    butterflies: a male-contributed antiaphrodisiac? Science
    193:419-420.
GILLIARD, E.T. 1969. Birds of Paradise and Bower Birds.
    New York: The Natural History Press of the Amer. Mus. Nat.
    Hist., 485 pp.
GWADZ, R.W. 1972. Neuro-hormonal regulation of sexual
    receptivity in female *Aedes aegypti*. J. Insect Physiol.
    18:259-266.
GWYNNE, D.T. 1981. Sexual difference theory: Mormon crickets
    show role reversal in mate choice. Science 213:779-780.
HALLIDAY, T.R. 1975a. An observational and experimental
    study of sexual behavior in the Smooth Newt, *Triturus vulgar*
    (Amphibia: Salamandridae). Anim. Behav. 23:291-322.
HALLIDAY, T.R. 1975b. On the biological significance of
    certain morphological characters in males of the Smooth
    Newt *Triturus vulgaris* and of the Palmate Newt *Triturus
    helveticus* (Urodela: Salamandridae). Zool. J. Linn. Soc.
    56:291-300.
HALLIDAY, T.R. 1976. The libidinous newt. An analysis of
    variations in the sexual behavior of the male Smooth Newt,
    Triturus vulgaris. Anim. Behav. 24:398-414.
HALLIDAY, T.R. 1977. The effect of experimental manipulations
    of breathing behavior on the sexual behavior of the Smooth
    Newt, *Triturus vulgaris*. Anim. Behav. 25:39-45.
HAPP, G.M. 1969. Multiple sex pheromones of the mealworm
    beetle, *Tenebrio molitor*. L. Nature 222:180-181.
HIRAI, K., H.H. SHOREY, and L.K. GASTON. 1978. Competition
    among courting male moths: male-to-male inhibitory pheromone
    Science 202:644-645.
HOUCK, L.D. 1980. Courtship behavior in the plethodontid
    salamander, *Desmognathus wrighti*. Amer. Zool. (Abs.)
    20:825.

HUMPHRIES, A.A., Jr. 1955. Observations on the mating behavior of normal and pituitary-implanted *Triturus viridescens*. Physiol. Zool. 28:73-79.

JACOBSON, M. 1965. Insect Sex Attractants. New York: John Wiley & Sons, Inc. 154 pp.

JORDAN, E.O. 1891. The Spermatophores of *Diemyctylus*. J. Morph. 5:263-270.

KIRKPATRICK, M. 1982. Sexual selection and the evolution of female choice. Evolution (in press).

KLEIMAN, D.G. 1974. Patterns of behaviour in histricomorph rodents. Pp. 171-209. *In:* The Biology of Histricomorph Rodents, I.W. Rowlands and B.J. Wier, eds. Symposia of the Zool. Soc. London, No. 34. New York: Academic Press. 482 pp.

LANDE, R. 1980. Sexual dimorphism, sexual selection, and adaptation in polygenic characters. Evol. 34:292-305.

LANDE, R. 1981. Models of speciation by sexual selection on polygenic traits. Proc. Natl. Acad. Sci. 78:3721-3725.

LANZA, B. 1959. Il corpo ghiandolare mento niero dei Plethodontidae (Amphibia, Caudata). Monitore Zoologica Italiano 67:15-53.

LEOPOLD, R.A., A.C. TERRANOVA, and E.M. SWILLEY. 1971. Mating refusal in *Musca domestica*: effects of repeated mating and decerebration upon frequency and duration of copulation. J. Exp. Zool. 176:353-360.

LLOYD, J.E. 1975. Aggressive mimicry in *Photouris* fireflies: signal repertoires by femmes fatales. Science 187:452-453.

MYERS, J. 1972. Pheromones and courtship behavior in butterflies. Am. Zool. 12:545-551.

MYERS, J., and L.P. BROWER. 1969. A behavioural analysis of the courtship pheromone receptors of the Queen butterfly, *Danaus gilippus berenice*. J. Insect Physiol. 15:2117-2130.

NELSON, K. 1964. Behavior and morphology in the glandulocaudine fishes (Ostarophysi, characidae). Univ. Calif. Publ. Zool. 75:59-152.

NOBLE, G.K. 1927. The plethodontid salamanders; some aspects of their evolution. Amer. Mus. Nov. 249:1-26.

NOBLE, G.K. 1931. The Biology of the Amphibia. New York: McGraw-Hill. 577 pp.

O'DONALD, P. 1977. Theoretical aspects of sexual selection. Theor. Pop. Biol. 12:298-334.

O'DONALD, P. 1980. Genetic Models of Sexual Selection. London: Cambridge University Press. 250 pp.

ORGAN, J.A. 1961. Studies on the local distribution, life history, and population dynamics of the salamander genus *Desmognathus* in Virginia. Ecol. Mongr. 31:189-200.

PERSOONS, C.J., and F.J. RITTER. 1979. Pheromones of cockroaches. Pp. 225-236. *In:* Chemical Ecology: Odour Communication in Animals, F.J. Ritter, ed. New York: Elsevier/North-Holland Biomedical Press. 427 pp.

PLISKE, T.E., and T. EISNER. 1969. Sex pheromone of the queen
    butterfly: biology. Science 164:1170-1172.
PLUMMER, E.L., T.E. STEWART, K. BYRNE, G.T. PEARCE, and R.M.
    SILVERSTEIN. 1976. Determination of the enantiomeric
    composition of several insect pheromone alcohols. J. Chem.
    Ecol. 2:307-331.
POOL, T.B., and J.N. DENT. 1977. The ultrastructure and the
    hormonal control of product synthesis in the hedonic glands
    of the red-spotted newt Notophthalmus viridescens. J. Exp.
    Zool. 201:177-202.
POOL, T.B., J.N. DENT, and K.KEMPHUES. 1977. Neural regulation
    of product discharge from the hedonic glands of the red-
    spotted newt, Notophthalmus viridescens. J. Exp. Zool.
    201:203-220.
PRIESNER, E. 1979. Specificity studies on pheromone receptors
    of noctuid and tortricid lepidoptera. Pp. 57-71. In:
    Chemical Ecology: Odour Communication in Animals, F.J.
    Ritter, ed. New York: Elsevier/North-Holland Biomedical
    Press. 427 pp.
RIEMANN, J.G., D.J. MOEN, and B.J. THORSON. 1967. Female
    monogamy and its control in houseflies. J. Insect Physiol.
    13:407-418.
RITTER, F.J. Ed. 1979. Chemical Ecology: Odour Communication
    in Animals. New York: Elsevier/North-Holland Biomedical
    Press. 427 pp.
ROGOFF, J.L. 1927. The hedonic glands of Triturus viridescens;
    a structural and functional study. Anat. Rec. 34:132-133.
ROTH, L.M., and G.P. DATEO. 1966. A sex pheromone produced by
    males of the cockroach Nauphoeta cinerea. J. Insect Physio
    12:255-265.
SALTHE, S.N. 1967. Courtship patterns and the phylogeny of the
    urodeles. Copeia 1967:100-117.
SEVER, D.M. 1976a. Morphology of the mental hedonic gland
    clusters of plethodontid salamanders (Amphibia, Urodela,
    Plethodontidae). J. Herpetol. 10:227-239.
SEVER, D.M. 1976b. Induction of secondary sexual characters
    in Eurycea quadridigitata. Copeia 1976:830-833.
SHOREY, H.H. 1973. Behavioral responses to insect pheromones.
    Ann. Rev. Ent. 18:349-380.
SIGNORET, J.P. 1976. Chemical communication and reproduction
    in domestic mammals. Pp. 243-256. In: Mammalian Olfaction,
    Reproductive Processes, and Behavior. R.L. Doty, ed.
    New York: Academic Press. 344 pp.
SIGNORET, J.P., and F. du MESNIL du BUISSON. 1961. Etude de
    comportement de la Truie en oestrus. Pp. 171-175. In: 4th
    Congres Inter. de Reprod. Animale et Insemination Artificiel
SVENSON, B.W. 1972. Flight period, ovarian maturation, and
    mating in Trichoptera at a south Swedish stream. Oikos
    23:370-383.

SWAILES, G.E. 1971. Reproductive behavior and effects of the male accessory gland substrate in the cabbage maggot, *Hylemya brassicae*. Ann. Entomol. Soc. Amer. 64:176-179.

THORNHILL, R.A. 1976a. Sexual selection and paternal investment in insects. Amer. Nat. 110:153-163.

THORNHILL, R.A. 1976b. Sexual selection and nuptial feeding behavior in *Bittacus apicalis* (Insecta: Mecoptera). Amer. Nat. 110:529-547.

TINBERGEN, N., B.J.D. MEEUSE, L.K. BOEREMA, and W.W. VAROSSIEAU. 1943. Die Balz des Samtmalters, *Eumenis* (=*Satyrus*) *semele* (L.). Z. Tierpsychol. 5:182-226.

TINBERGEN, N., and J.J. TER PELKWIJK. 1938. De Kleine Watersalamander. De Levende Natuur 43:232-237.

TUTTLE, M.D., and M.J. RYAN. (1981) Bat predation and the evolution of frog vocalizations in the neotropics. Science 214(4521):677-678.

TWITTY, V.C. 1955. Field experiments on the biology and genetic relationships of the California species of *Triturus*. J. Exp. Zool. 129(1):129-148.

TWITTY, V.C. 1961. Experiments on homing behavior and speciation in *Taricha*. Pp. 415-459. *In*: Vertebrate Speciation, W.F. Blair, ed. Austin: Univ. Texas Press. 642 pp.

WADE, M.J., and S.J. ARNOLD. 1980. The intensity of sexual selection in relation to male sexual behaviour, female choice, and sperm precedence. Anim. Behav. 28:446-461.

WEYGOLDT, P. 1966. Vergleichende Untersuchungen zur Fortpflanzungsbiologie der Pseudoscorpione: beobachtungen über das Verhalten, die Samenübertragungsweisen und die Spermatophoren einiger einheimischer Arten. Z. Morphol. Ökol. Tiere 56:39-92.

WEYGOLDT, P. 1969. The Biology of Pseudoscorpions. Cambridge, Mass.: Harvard Univ. Press. 145 pp.

ZELLER, E. 1890. Über die Befruchtung bei den Urodelen. Zeit. fur wiss. Zool. 49:583-601.

CHEMICAL SYSTEMATICS AND HUMAN POPULATIONS

Henry C. Harpending

Department of Anthropology
University of New Mexico
Albuquerque, New Mexico

Richard H. Ward

Department of Medical Genetics
University of British Columbia
Vancouver, Canada

Human populations offer many advantages for studying the processes generating and maintaining local genetic differentiation, that is, differentiation among neighboring demes or among regions within a single species. There are a large number of Mendelian biochemical traits or markers which vary in some or all human populations, and techniques for determining these are well standardized. In addition humans are unique because they can provide high quality information about their own family histories, reproductive histories, and other important variables. Our purposes in this paper are to emphasize the differences in purpose and procedure between local or intraspecific studies and studies of biochemical differentiation among species or higher taxa, to discuss some current work on human population structure, and to discuss the relevance of local differentiation to evolution.

SHORT AND LONG TERM PROCESSES

It is important to define and distinguish spatial and temporal scales in any discussion of biochemical variability.

The processes accounting for biochemical differences among
genera or higher taxa are very different from those accounting
for differences among villages in a valley.

At one extreme are studies of the evolution of certain
molecules in vertebrates, mammals, or some taxon higher than
the species (Nei 1975; Ayala this volume), evolution over time
scales of thousands to millions of years. The major issues in
molecular evolution concern the determinants of the rate of
amino acid substitution in peptides over evolutionary time
periods and of the level of polymorphism within species during
the course of evolution. This is the battleground of the
neutralism controversy; many biologists would account for most
molecular evolution and intraspecific polymorphism by the
directionless processes of mutation and genetic drift of
selectively neutral variants of ancestral molecules (Kimura
1968; King and Jukes 1969), while others believe that these
processes are driven by natural selection in most cases.

At the other extreme are studies of differences in the
frequency of biochemical markers among populations of a single
species. These local fluctuations are caused by genetic drift,
leading to continuous dispersal of gene frequencies each
generation, and gene flow among groups counteracting the
effects of genetic drift. Even though mutation and occasion-
ally natural selection may account for the macroscopic frequencies of molecular forms in a species, there is no reason
to expect that natural selection accounts for any substantial
part of the local fluctuations with which we are concerned.
Very strong selection in opposite directions in different
areas could conceivably cause local patterns such as clines,
but there are no very convincing examples in the literature
of human population genetics except at the spatial scale of
continents where traits like skin color, facial anatomy, and

other macroscopic features probably do reflect natural selection
and adaptation. Conversely there is almost no empirical nor
theoretical reason that local population structure observed
in human populations should have any effect on long term
molecular evolution.

In other words there is enough gene flow among human groups
at every level, and probably among populations of most mammals,
that the evolution of local biochemical differences and the
evolution of species-wide characteristics are nearly uncoupled
processes. The former is driven by local, short term genetic
drift and gene flow, the latter by long term genetic drift,
mutation, and natural selection. This conclusion depends upon
the presence of high enough levels of gene flow to render the
species effectively panmictic over evolutionary time periods.
Unless local isolation is extreme, long term drift of overall
frequencies within a species is effectively independent of
local structure. Weiss and Maruyama (1976) summarize the
evidence for high rates of interdeme gene flow in technologically
primitive human groups.

We have emphasized the distinction between long term and
short term evolution in order to clarify why anthropologists
and biologists should wish to pursue the study of population
structure. Since local structure is driven nearly exclusively
by migration and drift, it provides in effect an elegant
chemical assay of social organization. We can approach the
measurement of dispersal and mating distances, local deme size,
and other parameters simply by looking at tissues in a laboratory.

Figure 1 exemplifies the kind of data which we are discussing;
it shows the frequency of a particular kind of transferrin
in various groups of !Kung Bushmen of the northwest Kalahari
Desert. Typically we will have information like this from
five to several dozens of loci where more than one allele is

present at appreciable frequency. In addition there will often
appear at several of these loci very rare alleles present in
only a few of the groups. In this paper we discuss standard
approaches which work best with loci with several relatively

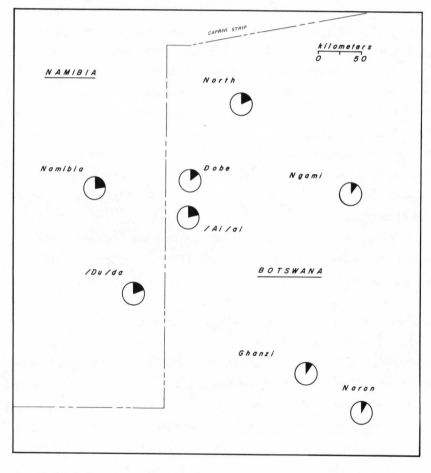

FIGURE 1.   Transferrin D frequencies in !Kung Bushman
             groups.

common alleles but do not gain much information from rare
variants. Slatkin (1980) has been developing very different
methods which seem to work best with rare genes.

MODELS OF LOCAL STRUCTURE

We summarize the spatial patterns of a large number of
loci by calculating for each allele a matrix of normalized gene
frequency covariances among groups and averaging these to
obtain a matrix R summarizing statistically the amount of local
gene frequency differentiation and the pattern of similarity or
relationship between groups (Harpending and Jenkins 1973). The
average of the diagonal elements of this matrix is Wright's F_{st}
statistic, a well-known measure of the overall dispersion of
local frequencies from the common mean. Individual diagonal
entries of this matrix can be interpreted as the genetic
distance of the corresponding group from the gene frequency
centroid of the region under study; that is to say, the
magnitude of these diagonal terms tells us how much local
variation there is in our sample.

Pairwise genetic distances are readily calculable, but it
is usually more convenient to plot the first several eigenvectors of R since these are "principal coordinates" and such a
plot provides a genetic "map" of similarity among groups which
can be compared to geography, to cultural or linguistic
indicators of proximity, and so forth. These vectors reveal
the patterning in space of local differentiation. Such maps
are imperfect representations of the information in R, but they
are optimal least squares projections of the intergroup distances
into a low number (usually two) of dimensions.

Figure 4 shows such a map of genetic distances among
several !Kung Bushman groups together with a geographic map
of their locations. We discuss below the interpretation of

discrepancies between these two maps. Our objective in
modelling processes leading to local genetic structure is to
predict the structure of the observed genetic covariance matrix
R from demographic information. We mention in the appendix a
tentative technique for the reverse operation, that is, for
inferring the pattern of gene flow in a region directly from
the gene frequency distributions.

We approach the problem first with an algebraic model
which is well known in the genetics literature (Malecot 1969;
Smith 1969; Bodmer and Cavalli-Sforza 1968). In this model
the groups or "islands" are populations within which random
mating occurs in discrete time. After reproduction, these
islands exchange a random sample of their members with each
other according to a migration matrix which is the backward
transition matrix of a reversible markov process. In addition
there is some exchange with one or more populations, called
"continents," outside the group under study. The gene
frequencies in these "continents" are assumed to be constant
over time, so that this immigration from continents acts as
a stabilizing pressure to prevent all the gene frequencies
from drifting to loss or fixation (see fig. 2). Continental
immigration is not absolutely necessary to the model, since
fixation or loss occurs over very long time periods, while we
are primarily interested here in the short-time patterning.

Figure 3 is a representation of frequencies of a single
gene through time in a hypothetical group of islands, each of
which receives a small amount of immigration from the continent
and each of which exchanges members with the others, so that
the gene frequencies of the islands remain clustered together
in time. The gene frequency on the continent (π) is the
expected value of these island gene frequencies. That is,
the mean of a large number of experiments such as this would

approach π but any particular experiment, such as the one we represent in fig. 3 or such as we encounter in field studies, will deviate in an unknown way. Confusion may arise because our model predicts mean squared deviations from the unknown mean π, but there is no way in real data to measure the dispersion about π.

If we have sampled gene frequencies in these islands at the point marked "present" in fig. 3, we compute normalized

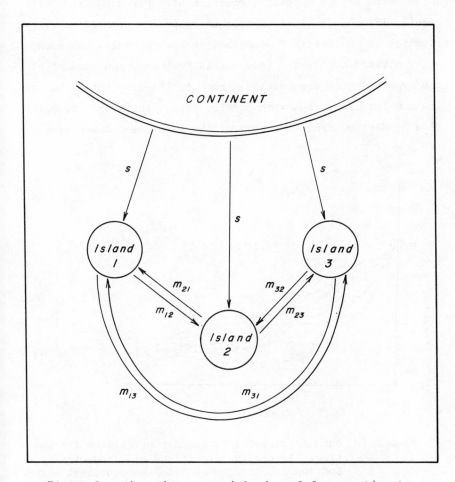

FIGURE 2. *Migration among islands and from continents.*

gene frequency variances and covariances R among groups
(normalized because they are divided by the mean times its
complement P(1-P), providing a statistic nearly independent of
the mean P). It is clear that our matrix of estimated covariances R measures only the dispersion around the mean P
of our sample.

With data about group sizes and intergroup migration it is
straightforward to compute a matrix of predicted genetic
covariances X (see appendix, equation 6). For comparison with
genetic information it is better to compute a related matrix
F, which is the matrix X described in the appendix with each
entry divided by $\pi(1-\pi)$; this matrix F is then independent of
the expected gene frequencies π both in its computation and in
its use for comparison with genetic data. The matrix X, which
is a prediction from migration of the normalized dispersion

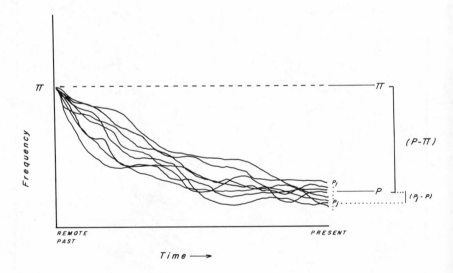

FIGURE 3.   Gene frequencies in a group of islands through
             time.   Initially each island had frequency π.
             Continuous immigration from the continent counter-
             acts drift and prevents fixation.

around π, will always differ a great deal from R.

We show in the appendix that for this special case of a
single continent the matrix R is essentially the matrix F less
the leading entry of its spectral decomposition. That is, the
matrix constructed from the first eigenvalue of F and the outer
product of its leading right and left eigenvectors (the first
dyad) predicts the quantity $\frac{(P-\pi)^2}{P(1-P)}$ and so adds a constant
quantity to each entry of F, while the remaining dimensions
of F describe pattern within the group of islands and are the
prediction of the matrix R computed from gene frequency data.
This leading dyad of F, which predicts the dispersion of the
island mean from the continental mean, is a matrix in which
all entries are the number

$$\overline{f} \cong 1/(4Ns + 1)$$

where N is the total population of the group of islands and s
is the rate of immigration from the continent. Not a few studies
have compared gene frequency variance among populations with the
quantity $1/(4Ns + 1)$ estimated in some way; this quantity (see
fig. 3) is precisely what cannot be observed in gene frequencies,
since only the dispersal among islands away from P, rather than
the dispersal around π can be estimated from data.

In this model, when each island receives the same proportion
of immigrants from the continent, the unobservable leading dyad
of the genetic covariance matrix F corresponds to the leading
dyad of the migration matrix. This leading dyad of a stochastic
matrix is simply a matrix of which each column is the expected
proportion of the total population residing on the island
corresponding to that column. Any patterning in interisland
migration is in the second and higher dyads of the migration
matrix, and this patterning in migration leads to the spatial
patterning in gene frequencies which we study in R. Details are

in the appendix.

This model is the usual starting point for studying spatial pattern. We are interested in whether the detailed correspondence between migration and covariance present in the model is present in data from human and nonhuman populations; the answers to this question in the literature tend to be equivocal. It is ordinarily observed that both migration intensity and genetic similarity decline with distance. In figure 4, the Bushman data plots of the leading eigenvectors of the matrix R, and the second and third (the first has no pattern) normalized eigenvectors of the migration matrix show an additional kind of agreement. The group around Lake Ngami are in fact employees of cattle keepers whose relatives and social ties are all with the Xangwa and Kai Kai regions to the west. The vectors both of the genetic covariance matrix R and of the migration matrix M place this group with their relatives.

This item of sociological information is readily apparent from the genetic structure, while other sociological information is falsified by the genetics, as for example in the sharp division between the !Kung of Ghanziland and those to the north. Such detail is fascinating to anyone with an interest in !Kung, and there are many examples in the literature showing similar correspondences between genetic structure and social structure. These methods should become especially useful in work with other animal and plant species where the social data are not so simply obtained by interviewing subjects. For example, genetic structure could provide a simple means for measuring gene dispersal in

----

FIGURE 4.   *Distances among !Kung Bushman groups reconstructed from marker gene frequencies and from intergroup migration. Actual distances are shown in figure one.*

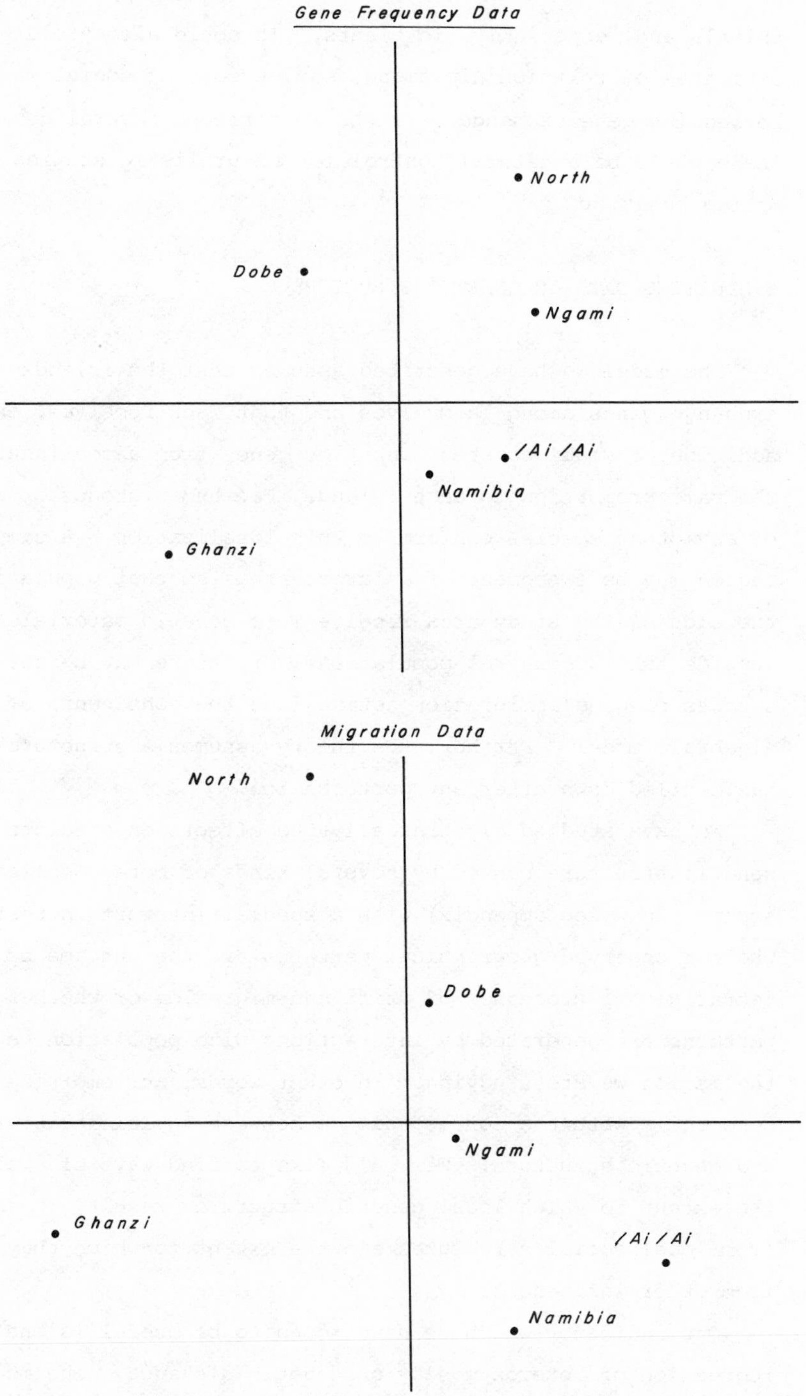

Gene Frequency Data

• North

Dobe •

• Ngami

• /Ai /Ai

• Namibia

• Ghanzi

Migration Data

North •

• Dobe

• Ngami

Ghanzi •

• /Ai /Ai

• Namibia

DISTANCES AMONG !KUNG BUSHMAN GROUPS

animals and, especially, in plants. It could also yield
estimates of relationship among, say, troops of social mammals,
reflecting gene exchange over the most recent several generations
These would be a natural control for naturalistic studies of
mating behavior.

EVALUATING DATA ON GENETIC STRUCTURE

The model we have described assumes that the islands
exchange genes among themselves and that each receives, in
addition, a small constant input of genes from a continent,
the same proportion to each island. Few human groups or groups
of any other species conform to this idealization. A sampled
region may be a segment of a larger area, so that populations on
the edge of the study area receive more genetic material from
outside than do central populations, or, there may be several
sources of gene immigration acting like the continents of our
algebraic model. Further, the theory assumes a structure which
has settled down after any perturbation.

We have studied algebraically the effects on predicted
genetic structure caused by several kinds of more complicated
assumptions (see appendix) with a special interest in testing
whether observed geographical patterns are the outcome of
intraregional processes of drift and migration or whether these
patterns are generated by interactions with populations outside
the region we are studying. In other words, our empirical
concern is with the correspondence between social structure
and genetic structure. We would like to find ways of evaluating
the extent to which local genetic structures result, in fact,
from local social structures and the extent to which they result
from other influences.

One indicator which we have found to be useful is the
regression of heterozygosity on genetic distance. The model

we have discussed can be derived by reasoning about either gene
frequency covariances or about the probability of identity of
genes in the same and different islands. Pursuing this, it is
easy to show (see appendix) that the genetic distance of an
island from the gene frequency centroid (the overall mean gene
frequencies of the population) and the relative homozygosity of
that island should be linearly related if exchange with
populations outside the region is the same for each island. If
migration declines monotonically with distance, islands on the
edge of the collection should show the largest genetic distance
from the centroid, and they should lose genetic variability
the most rapidly because they are peripheral. More precisely,
the average heterozygosity of the i'th island should equal the
heterozygosity of the overall population mean gene frequencies
multiplied by $(1-r_{ii})$ where r_{ii} is the genetic distance of the
island from the centroid. If gene flow from outside the region
varies in amount from island to island, this linear relationship
no longer holds. Very isolated islands should be less heterozygous than the linear prediction, while islands which receive
more genes from the outside should be more heterozygous. Thus
the theory indicates that we might gain a great deal of useful
insight by examining this regression, paying special attention
to outliers. It is surprising that deviations from the
regression depend upon the amount of gene flow from the outside
but not upon the number of sources. The linear form does
continue to hold, for example, if each island exchanges with a
different continent.

We have examined the regression of heterozygosity on
distance from centroid in six human population samples from
the literature. The islands in two of these samples are large
tribal groups distributed over continental areas -- Central and
South America and southern Africa -- while in the remaining four

samples the islands are on the scale of local demes.

Figure 5 shows a scattergram of heterozygosity versus
distance for twenty-one American Indian tribes of Central
America and the northern half of South America (Fitch and Neel
1969). The predicted regression line is shown on the figure.
The fitted regression is close to the theoretical one, and the
correlation between heterozygosity and distance from centroid
is -.58. We are not so much concerned with the fit of the
regression, however, as with the interpretation of outliers.
In this sample the outliers which our theory indicates to be
isolated from extraregional gene flow are numbered 7 and 20.
These are respectively the Guaymi of Panama, a small tribe
on the northern edge of our sample, and the Yanomamo of
Amazonia, one of the most isolated of South American tribes
today. For these groups continental immigration corresponds
primarily to Spanish and Portuguese genes. Little is known
about the Guaymi, but it is known that the Yanomamo have
experienced little or no European admixture, so their position
as outliers is consistent with our previous expectation. The
Shipibo of eastern Peru, group 14, show excess heterozygosity.
Their identity and ethnography are poorly known; our theory
indicates that they are more admixed than other South American
tribes.

Figure 6 shows heterozygosity and distance for fifteen
populations of various size and cultural complexity from
southern Africa (Harpending and Jenkins 1973, Harpending and
Chasko 1976). There is clustering around the theoretical
regression line even though the groups range in size from
several hundreds of hunter-gatherers from a valley in the
Kalahari Desert to samples of Bantu-speaking tribes of several
hundreds of thousand people. Group 5 is a low outlier; these
are the Naron Bushman of the Western Kalahari around Ghanzi,

Botswana.   They are culturally and genetically isolated even
from their neighbors the !Kung, and they seem to be declining
in size.   On the other hand groups 13, 14, and 15 are high
outliers.   These are the so-called coloured populations, and
they have indeed experienced more extra African gene input
than any of the other groups.   They are descendants in varying
proportions of several major African, Asian, and European
populations.   Group 14, the Johannesburg Coloured, is the
furthest from the theoretical regression, and it is the most
urbanized and cosmopolitan of the Coloured groups.   Groups 13

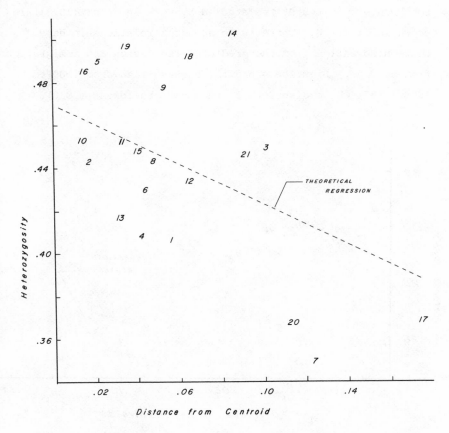

FIGURE 5.   Heterozygosity and distance   in South American
            tribes.

and 15 are from Kuboes and Sesfontein in Namibia where they
have experienced less extra African immigration. The correspondence in southern Africa between the genetic theory and
the ethnographic information is excellent.

Figure 7 shows seven populations from the Åland islands
between Finland and Sweden (Jorde, 1979). Group three, Foglo,
is the only clear outlier. Of all these groups the genetic
distance between Foglo and Finland is the smallest, indicating
more gene flow into Foglo than into the others. Foglo is also
directly on a shipping lane between Finland and Sweden.

Ten districts of the Papago Indian reservation in the
southwestern United States studied by Workman *et al* (1973) are
shown in figure 8. There is no apparent relationship here
between deviation from the predicted regression and immigration
from outside. District 4 had the highest rate from 1900 to
1959 (.187) while districts 3 and 9 had the lowest (.030).

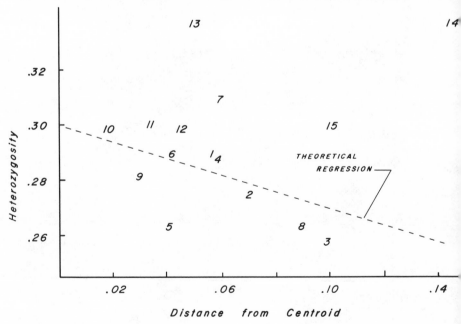

FIGURE 6.   *Heterozygosity and distance in Southern African*
            groups.

Workman *et al* state that their figures for immigration mostly
reflect Papago moving locally across the reservation boundary
and do not reflect long-range immigration. The fitted
regression line is very close to the predicted line.

Figure 9 shows heterozygosity and distance for fifty
villages of Yanomamo Indians of the Amazon Basin. This is
an expanding and very isolated tribe of hunters and gardeners
(Weitkamp *et al* 1972a,b; Ward 1972; Ward *et al* 1975; Tanis
et al 1973; Gershowitz *et al* 1972). The Yanomamo are noted
for their very high levels of genetic differentiation among
villages and for their mechanism for internal gene flow.
Whole segments of villages, usually kin groups, sporadically
establish new villages or merge with some other group. The
fitted regression here has a positive rather than a negative
slope; there is highest heterozygosity at the edges rather
than at the center of the Yanomamo. This phenomenon raises
doubts about the extent to which gene frequency diversity
among Yanomamo villages is generated by the population structure
within the region; these data indicate that a significant
part of the diversity is due to differential admixture from
surrounding tribes. The extreme outliers are villages 47
and 48 (labelled 15L and 15O in the cited original publications).
These are both geographically peripheral to the east and are
probably highly admixed with Arawak-speaking groups.

Finally, figure 10 shows eight groups of !Kung Bushmen
hunter-gatherers from the northwest Kalahari Desert of Botswana
and Namibia (Harpending and Jenkins 1974). Groups 6 and 4,
/Du /da and Namibia, are low outliers. The former group is
in otherwise uninhabited desert far from any other people,
while the Namibian population is isolated from other groups
by the Namibian government. Continental immigration here
corresponds to genes from Bantu-speaking peoples who are

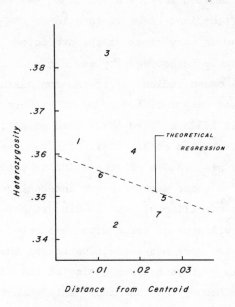

FIGURE 7.   Heterozygosity and distance in Åland Island
            parishes.

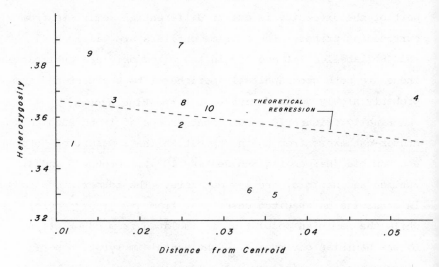

FIGURE 8.   Heterozygosity and distance in Papago districts.

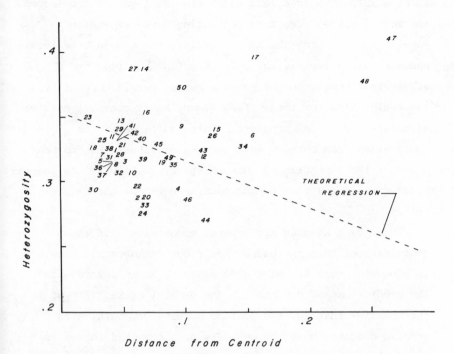

FIGURE 9.   Heterozygosity and distance in Yanomamo
            villages.

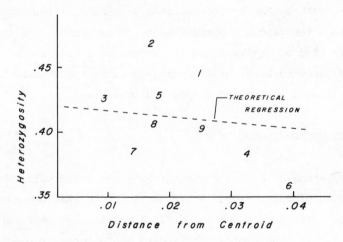

FIGURE 10.   Heterozygosity and distance in !Kung
             Bushman regions.

pushing into this area, and these two !Kung groups are indeed
the most isolated from this continuing Bantu expansion.
Population 2 is from the Xangwa Valley, the site of the local
administrative center, the only store in the area, and a
substantial population of Bantu-speaking people. The fitted
regression line for these !Kung groups has a steeper negative
slope than our model predicts. This indicates excess
immigration from outside into the center rather than to the
edges of the !Kung range, conforming very well with the central
placement and the ties to the outside world of the Xangwa
Valley.

These case studies are hopeful endorsements of the
usefulness of this approach. Where our ethnographic knowledge
is adequate, outliers make good sense to us in accord with
the predictions of our model. The general overall fit of
the model -- that is, the decline of heterozygosity with
genetic distance once the outliers are accounted for -- is
a reassuring indication that these regional gene frequency
distributions are largely generated by local population
structure rather than by interactions with populations outside
the region, the Yanomamo perhaps being an exception. We hope
that this kind of analysis and its refinements will become a
useful adjunct to behavioral studies in other species where
the social data are not so simply had by asking.

IMPLICATIONS FOR EVOLUTION

These kinds of data and methods have clear and well-
defined sociological uses. It is not so clear what consequences
population structure and local differentiation may have for
evolution. In other words, why should evolutionary biologists
be interested in this material? There are several possibilities

none of which seem to be well documented or studied in natural populations. First, the amount of interpopulation gene flow can influence the extent to which local populations can adapt to the local environment. Restricted migration may allow local gene frequencies to track local aspects of the environment, while high levels of migration may swamp local fine tuning. Second, population structure may generate linkage disequilibrium favoring evolution of quantitative traits, that is, traits influenced by a large number of loci (Wright, 1931). Finally, population structure should be critical in the evolution of certain social and related kinds of genetic traits. The relationship between gene flow and adaptation to local environmental conditions has been thoroughly discussed in the literature (Endler 1977), the effect on linkage disequilibrium has been studied in a few papers and its consequences for evolution of quantitative traits need to be worked out in more detail. We wish to discuss here the potential uses of population structure data for the study of social evolution.

Social traits affect the fitness of individuals who carry them as well as the fitness of others with whom interaction occurs. Examples include traits affecting behavioral interactions as well as others not usually thought of this way. For example, immunity to an infectious disease affects both the fitness of the bearer of the immunity and of others of the same group. In human groups, mental illness may reduce the fitness of relatives if they are avoided as mates by others (Yokoyama 1980). Classical population genetics assigns fitnesses to individuals and then deduces consequences for population gene frequencies. This approach doesn't work for social traits, since the fitness of individuals depends both upon the genes they carry and on the genes of conspecifics with whom they interact.

The most familiar approach to modelling social traits is that of Hamilton (1964), who used the idea of the coefficient of relationship to simplify the algebra of social evolution. His essential finding is that the fitness of genes carried by an individual is increased when the individual helps others who are more closely related to him than the average and who interferes with or decreases the fitness of others less closely related to him than the average.

How does interaction occur with individuals more or less related than the average? There are three likely possibilities. First, pedigree relatives can be recognized and remembered, and ethologists seem to be finding much pedigree kinship in the social organization of mammals. Second, individuals may recognize aspects of the phenotypes of others and adjust their responses accordingly. Third, interaction within differentiated subpopulations always leads to interaction with others more closely related than the average. This relationship within subdivisions is raised by an amount F, where F is the statistic describing population differentiation which we have used in this paper.

These mechanisms often overlap. For example troops of many primate species are composed of one or more core matrilineages, so that within troops females especially tend to be genetically closely related. Members of troops in some species share more or less distinctive markings. Regardless of the mechanisms by which social behavior is cued, the statistical measures of subdivision that we have discussed above are appropriate parameters for modelling. Relatedness ultimately means shared genetic material and genetic correlation, while the cue systems which have evolved to detect relationship may be diverse.

For example, imagine two groups of equal size in which

the frequency of some allele is 80% in one and 40% in the
other. If interactions between individuals are random within
groups, then bearers of the allele interact with each other
$\frac{(0.8)^2 + (0.4)^2}{2}$ = 40% or 40% of the time. If this were a single
randomly mating group, the frequency of the allele would be
60% and interactions with others of the same type would occur
only 36% of the time. The difference between 40% and 36% is
not large, but such small differences can lead over time to
appreciable evolutionary consequences if the allele leads to
helping or harmful behavior. The genetic subdivision in this
example would correspond to pedigree relatedness within groups;
our point is that evolution acts on the gene identity and not
the geneological relatedness per se. In general, subdivision
should foster helping, cooperation, and altruism within groups,
and it should hinder the evolution of selfish behavior.

Social traits may turn out to be more amenable to comparative empirical investigation than traits which are
adaptations to the nonsocial environment. Some evolutionary
biologists believe that macroevolution is to be understood as
a series of unique appearances of strongly altered forms which
quickly (in evolutionary time) displace the preexisting type
(Stanley 1979). This new saltationism would relegate the
gradual change in gene frequencies as described in genetics
texts to fine tuning of traits and to adaptation to changes
which occur on a fast time scale. If this viewpoint is correct,
then two categories of phenomena where gradual gene frequency
change should be detectable and important are social behavior
and adaptation to infectious disease. Social behavior should
change rapidly in mammals, in large part because of the
importance of learning processes during ontogeny interacting
with genetic propensities. Infectious diseases can change
rapidly because the surface chemistry of many pathogens is

subject to mutation and fast evolution, so there is strong selective advantage for the ability to combat antigenic novelty. In order to understand the evolution of both of these phenomena, either theoretically or empirically, a knowledge of population structure is necessary. Social inter- actions occur largely within restricted groups and, by definitic affect the fitness of more than one individual; population structure interacts with infectious disease both because it affects disease transmission directly and because the immune response of an individual is a social act, determining his likelihood of retransmission of the disease to other members of the subpopulation.

Population structure also needs to be thoroughly considered when attempting to correlate gene frequencies with putative selective agents of the environment. Relatedness within groups corresponds to the same correlations among allelic frequencies among groups. Hence, all sorts of clines and other spatial patterns are generated simply by restricted dispersal, and these could correlate spuriously with features of the environment.

ACKNOWLEDGEMENTS

This work was supported in part by NIH grant GMO 7661 to Richard Griego, Department of Mathematics, University of New Mexico. We are grateful to Lisa Sattenspiel and Alan Rogers for helpful comments.

APPENDIX

   *This appendix contains algebraic justification for our
statements about genetic distance and heterozygosity. Here
we will:*

   a. *derive an expression for the genetic covariance matrix
       at equilibrium under the assumptions of equal
       stabilizing pressure to each island from a single
       continent;*
   b. *derive the parameters of the regression of hetero-
       zygosity on genetic distance from centroid for this
       case;*
   c. *remark that the genetic distance matrix converges in
       only a few generations for data typical of regional
       human populations;*
   d. *remark that if endogamy rates are greater than 50%, the
       genetic covariances uniquely determine the migration
       matrix;*
   e. *show that variation among islands in the proportion of
       continental immigrants distorts relative heterozygosity
       but leaves genetic distances almost unchanged, so that
       the regression in (b) does not hold;*
   f. *show that when each island receives the same proportion
       of immigrants in each generation, each from a different
       source continent, the genetic covariance matrix may be
       written*

$$X = X_{mig} + X_{drift}$$

   *where X_{mig} is the deterministic part of the dispersion
   about the immigrant mean, while X_{drift} is nearly the
   same as the covariance matrix in (a). In this case the
   regression in (b) still holds under certain conditions
   on the gene frequencies of the continents.*

   *We assume that we are studying a finite group of t sub-
populations (islands) which exchange, each generation, a
constant proportion of migrants with one or more external source
populations (continents) with unchanging gene frequencies at a
number of loci. Selection and mutation are assumed to be of
negligible effect, and migration among islands is a discrete
reversible Markov process.*
   *We will derive here approximations to the expected covariance
matrix of gene frequencies among islands under two special cases
of this model: (1) The case in which the rate of exchange with
a single continent varies from island to island; and (2) The*

case in which each island exchanges with a different continent
at a rate constant from island to island.

We emphasize that we are deriving expected values of random
variables; it would be possible to estimate higher moments by
the same methods for exploring statistical questions. We do
not pursue such questions here, and we assume that we have
perfect knowledge of observable quantities. We will also
present possibilities that we believe to be useful for inference
about population structure.

Standard Model

The model of equal input to islands from a single continent
is well known in population genetics (Bodmer and Cavalli-Sforza
1968; Smith 1969; Malecot 1969; Courgeau 1974; Carmelli and
Cavalli-Sforza 1976). We derive some of the properties of this
situation here, discuss how the theoretical quantities are
related to genetic statistics, and then show how the modificatio
we are studying perturb this model.

Migration among islands is described by an irreducible
symmetric matrix Z where Z_{ij} (=Z_{ji}) is the expected number of
migrants from the i'th to the j'th island each generation. The
symmetry of Z is an ordinary and reasonable assumption, implying
that there is no "tendency" toward cylical exchange. The total
population of the group of islands, N, is the sum of all the
entries of Z, while the row (or column) sums of Z give the
sizes of each island. The size of the i'th island divided by
the total N is the weight w_i of island labelled by i. We will
use w as a (column) vector of these weights, while the matrix
W is a diagonal matrix with w_i in the i'th diagonal position.

The backward transition matrix M* whose i, j'th entry is the
proportion of the population of island j which comes from island
i each generation is

$$M^* = Z \ W^{-1} x \ (\tfrac{1}{N})$$

The forward transition matrix is the transpose of the backward
matrix (since the process is reversible; Kemeny and Snell 1960)
and is

$$M^{*T} = W^{-1} Z x \ (\tfrac{1}{N}) = W^{-1} M^* W \tag{1}$$

We assume for convenience that M* has t distinct eigenvalues
(this will almost always be true for real data; it is not true
for certain symmetric models, e.g., Maruyama 1977), and that it
can be written

$$M^* = U^* \ \lambda^* \ V^{*T}$$

where U and V* are the right and left eigenvectors, respec-
tively. These vectors are unique up to a multiplicative
constant, that is if U* is a matrix of eigenvectors in columns,
so is*

$$U = U* \ F$$

if F is a positive diagonal matrix. Since U and its inverse
V* are arbitrary, depending, for example, on the way in which
a computer routine normalizes eigenvectors, we wish to rescale
them for convenience using the reversibility of M*. From
equation (1)*

$$M*^T = V* \ \lambda* \ U*^T = W^{-1} \ U* \ \lambda* \ V*^T \ W \tag{2}$$

If we choose F so that

$$V*F = W^{-1} \ U* \ F^{-1}$$

giving

$$F^2 = U*^T \ W^{-1} \ U* \qquad \text{(since } U*^T \ V* = I)$$

and write

$$V = V* \ F \qquad and \qquad U = U* \ F^{-1}$$

then

$$M*^T = V \ \lambda* \ U^T = W^{-1} \ U \ \lambda* \ V^T \ W$$

and

$$V = W^{-1}U \qquad or \qquad VV^T = W^{-1}$$

$$U^T = V^T W \qquad or \qquad UU^T = W \tag{3}$$

A standard property of Markov chains and this rescaling yield

$$U_{i,1} = w_i$$

$$V_{i,1} = 1 \qquad for \ all \ i$$

*If p is a vector of island gene frequencies for an allele
at some locus, then the expected value of this vector follows
at equilibrium*

$$E(p) = M*^T \ E(p) + S \ [\pi - E(p)] \tag{4}$$

where S *is a diagonal matrix with* s, *the proportion of each island's population exchanged with the continent per generation, in each diagonal position,* π *is the allele frequency on the continent, and* $[\pi - E(p)]$ *is a vector whose i'th entry is* $[\pi - E(p_i)]$. *We will write*

$$M = M* - S$$

and

$$E(p) = \bar{p}$$

so that equation (4) becomes

$$\bar{p} = M^T\bar{p} + S\pi \tag{5}$$

The *eigenvectors of* M *are the same as those of* M*, *while the eigenvalues are*

$$\lambda_k = \lambda_k^* - S$$

and in particular λ_1 = 1-s.   Hence *the solution of (5) is*

$$\bar{p} = (I - M^T)^{-1} S\pi$$

$$= \sum_k V_k U_k^T [1/(1 - \lambda_k)] S\pi$$

$$= V_1 U_1^T (1/s) S\pi$$

$$= \pi$$

since the eigenvectors U_i *are, except for the first, orthogonal to the constant vector* $S\pi$.

In *order to derive the covariance matrix of gene frequencies among islands, we write the gene frequency vector* p *as*

$$p = \pi + \Delta$$

so the column vector Δ *gives the deviations of the islands' gene frequencies from the mean* π. *The covariance matrix* X *is now simply the expectation of the outer product* $\Delta\Delta^T$.

Let p(t) *be the gene frequencies at time* t *before reproduction and migration occur. After reproduction this is transformed into*

$$p'(t) = p(t) + \epsilon(t)$$

where $\varepsilon(t)$ *is the vector of changes due to genetic drift, that is*

$$E(\varepsilon) = 0$$

$$E(\varepsilon\Delta^T) = 0$$

$$E(\varepsilon\varepsilon^T) = D^*$$

where D^* *is a diagonal matrix with i'th diagonal entry*

$$D^*_{ii} = \frac{E[p_i(1-p_i)]}{2W_i N} = \frac{\pi(1-\pi) - X_{ii}}{2w_i N}$$

Migration now occurs and the gene frequency vector becomes

$$p(t+1) = M^T[P(t) + \varepsilon(t)] + S\pi$$

so that

$$\Delta(t+1) = M^T[\Delta(t) + \varepsilon(t)]$$

and the covariance matrix follows

$$X(t+1) = E[M^T(\Delta(t) + \varepsilon(t))][(\Delta(t) + \varepsilon(t)) M^T]^T$$
$$= M^T[X(t) + D^*(t)] M$$

and at equilibrium

$$X = M^T X M + M^T D^* M \tag{6}$$

(Malecot 1969, Courgeau 1974).

We now introduce an approximation by substituting D for D^*, where D is also a diagonal matrix with

$$D_{ii} = \frac{\pi(1-\pi) - X_0}{2w_i N} \qquad X_0 = \sum_k w_k X_{kk}$$

that is, we substitute the average diagonal of the matrix, X_0*, for the value* X_{ii} *in the i'th diagonal position. This is justified when the variation among islands in* X_{ii} *is small; in numerical experiments with data from human populations the approximation made little difference in the answer.*

Equation (6) can be written in a different form, viz

$$\underset{\sim}{X} = \underset{\sim}{M}^T \underset{\sim}{X} + \underset{\sim}{M}^T \underset{\sim}{D} \tag{7}$$

in which $\underset{\sim}{X}$ is a column vector with t^2 elements which are the entries of X in dictionary order, $\underset{\sim}{D}$ is a column vector with the t^2 entries of D in dictionary order, and M^T is the kronecker product of M^T with itself, that is it is a t^2 by t^2 square matrix, the first t rows and t columns contain the entries of M^T each multiplied by M_{11}, the second t columns and first t rows contain the entries of M^T each multiplied by M_{21}, and so forth. This matrix has t^2 eigenvalues of $\lambda_j \lambda_k$ with corresponding right eigenvectors $V_j \otimes V_k$; the Kronecker product of the vectors V_j and V_k. Thus the solution of equation (7) is

$$\underset{\sim}{X} = (\underset{\sim}{I} - \underset{\sim}{M}^T)^{-1} \underset{\sim}{M}^T \underset{\sim}{D}$$

with dyad representation

$$\underset{\sim}{X} = \sum_{j,k} \frac{\lambda_j \lambda_k}{1-\lambda_j \lambda_k} \quad (V_j \otimes V_k)(U_j \otimes U_k)^T \underset{\sim}{D} \tag{8}$$

This will be more tractable if we write out the equation for a specific entry of $\underset{\sim}{X}$, for example the element corresponding to X_{ij} of X (This is the $[i(t-1)+j]$'th entry of $\underset{\sim}{X}$). From equation (8)

$$X_{ij} = \sum_{k,\ell} \left(\frac{\lambda_k \lambda_1}{1-\lambda_k \lambda_\ell}\right) \quad V_{i\ell} V_{jk} \left\{\sum_m U_{m\ell} U_{mk} D_{mm}\right\} \tag{9}$$

The summation { } over the index m is

$$\sum_m U_{mk} U_{m\ell} w_m^{-1} \left(\frac{\pi(1-\pi)-X_0}{2N}\right)$$

but this is

$$(U^T W^{-1} U)_{k\ell} \times \left(\frac{\pi(1-\pi)-X_0}{2N}\right)$$

$$\left(\frac{\pi(1-\pi)-X_0}{2N}\right) \times \left\{\begin{matrix}1 \text{ if } k = \ell \\ 0 \text{ if } k \neq \ell\end{matrix}\right\}$$

since $U^T W^{-1} U = V^T U = I$ by equations (3). Then equation (9) becomes

$$X_{ij} = \sum_k B_k V_{ik} V_{jk}$$

that is

$$X = VBV^T \tag{10}$$

where B *is a diagonal matrix with*

$$B_{kk} = (\frac{\lambda_k^2}{1-\lambda_k^2}) \ (\frac{\pi(1-\pi)-X_0}{2N})$$

This is not quite a spectral decomposition of X *since the matrix* V *is not orthogonal. It does furnish such a decomposition of the weighted covariance matrix* $W^{\frac{1}{2}}XW^{\frac{1}{2}}$ *since the matrix* $W^{\frac{1}{2}}V$ *is orthogonal. Such a weighted matrix may be, at any rate, the more natural object for the study of population structure. If the islands are all of the same size then* $U \propto V$ *and these are eigenvectors of* X.

If the endogamy rates, that is the diagonal entries of M, *are all greater than one half, then Gershgorin's theorem guarantees that all the eigenvalues* λ_k *of* M *are positive and equation (10) defined a one to one relationship between gene-frequency covariances and the migration matrix. It is possible from this to infer migration rates from the gene frequency distributions.*

Interpretations

We will use two interpretations of the elements of the matrix X.

First, X_{ij} *is the expected value of the covariance between gene frequencies of islands* i *and* j, *i.e.*

$$E \ (p_i-\pi) \ (p_j-\pi) = X_{ij}$$

This covariance matrix is often normalized as

$$f_{ij} = \frac{X_{ij}}{\pi(1-\pi)} \ .$$

This matrix F *is the Wahlund covariance matrix.*

Second, the diagonal entries of X *are the expected decrements in heterozygosity from the heterozygosity of the continent, that is if* H_i *is the heterozygosity of island* i *averaged over many loci and if* H_π *is the same average for the continental gene frequencies, then*

$$E(H_i) = H_\pi - X_{ii} \tag{11}$$

To see this, write the heterozygosity for a single allele on island i as

$$P_i(1-p_i) = (\pi+\Delta_i)(1-\pi-\Delta_i) = H_\pi + (1-2\pi)\Delta_i - \Delta_i^2$$

For this standard model $E(\Delta_i)$ is equal to zero and $E(\Delta_i^2)=X_{ii}$, so equation (11) holds. Below we will consider the case of multiple continents differing in gene frequency, and equation (11) will be valid if, over many alleles and loci, the average product of the deviation of the grand immigrant mean from one half (the term $1-2\pi$) and the deviation of the mean of each island i (the term Δ_i) disappears. In the case of multiple continents, this deviation has a deterministic component. The product does not disappear if for some continent the frequencies of common genes are regularly closer to one and the frequencies of uncommon genes regularly closer to zero than on other continents. This condition is not the same as the condition that each continent have the same average heterozygosity; consider a "continent" which is a mixture of two others. By Wahlund's principle the hybrid continent would have higher mean heterozygosity than the average of the parents, but the product of deviations would still disappear if it disappeared for each parent.

Statistics

When we are considering data on gene frequencies in a group of subpopulations, there is no information at all ordinarily about the expectation π. For example genetic covariances computed from data yield moments around $\hat{p}$, the sample mean, rather than moments around π. We discuss here some properties of the sample covariance matrix R which has entries

$$R_{ij} = \sum_{\substack{alleles \\ loci}} \frac{(p_i-\hat{p})(p_j-\hat{p})}{\hat{p}(1-\hat{p})} \Big/ \# \text{ of alleles}$$

where $\hat{p} = \sum_k w_k p_k$. To first order the expected value of this statistic is

(Harpending and Jenkins, 1973; Workman et al. 1973)

$$E(R_{ij}) = \frac{X_{ij}-\bar{X}_i-\bar{X}_j+\bar{X}}{\pi(1-\pi)-\bar{X}} \tag{12a}$$

where $\bar{X}_i$ *is the column or row average*

$$\bar{X}_i = \sum_k w_k X_{ik}$$

and $\bar{X}$ *is the overall weighted average value of the entries of X.*

Equation (12a) *shows that sample R statistics are proportional to the entries of the matrix X "centered" to zero, that is to the entries of X less the column mean less the row mean plus the grand mean.* Equation (7) *is a statement of the equilibrium form of the covariance matrix X. The process of convergence follows*

$$[\underset{\sim}{X} + \delta\underset{\sim}{X}(t+1)] = \underset{\sim}{M}^T[\underset{\sim}{X} + \delta\underset{\sim}{X}(t)] + \underset{\sim}{M}^T\underset{\sim}{D}$$

so that

$$\delta\underset{\sim}{X}(t+1) = \underset{\sim}{M}^T[\delta\underset{\sim}{X}(t)]$$

where $\delta\underset{\sim}{X}(t)$ *is the deviation from equilibrium at generation t. The Kronecker matrix* $\underset{\sim}{M}^T$ *has* t^2 *dyads with eigenvectors* $\lambda_j\lambda_k$ *corresponding to all* t^2 *pairings of the dimensions of* $\underset{\sim}{M}^T$. *If this equation is written out, it is apparent that the dyad operator with coefficient* λ_1^2 *changes all entries of X by a constant value and the* $(2t-2)$ *terms with coefficient* $\lambda_1\lambda_J$ $(j \neq 1)$ *change entries of* $\underset{\sim}{X}$ *corresponding to single rows and columns of X by a constant amount. Hence these terms do not change the entries of the expected matrix of sample covariances R nor the genetic distances. Thus the rate of convergence of the matrix R is determined by the value of* λ_2^2 *which is* $(\lambda_2^* - s)^2$ *where* λ_2^* *is the second eigenvalue of the migration matrix M*. This means that the convergence of the dispersion about the sample mean may ordinarily be very rapid, on the order of several generations, while the equilibrium dispersion of island mean gene frequencies* $\hat{p}$ *about* π *may happen much more slowly. For a numerical computation showing this effect see* Workman et al. *(1973).*

The expectation of the i'th diagonal entry of R is

$$E(R_{ii}) = \frac{X_{ii} - 2\bar{X}_i + \bar{X}}{\pi(1 - \pi) - \bar{X}}$$

This is interpretable as the genetic distance of the i'th island from the centroid of the island gene frequencies if distance is the simple geometric distance of Harpending and Jenkins (1974), viz

$$d_{ij} = R_{ii} + R_{jj} - 2R_{ij}$$

For the standard model with one continent and equal exchanges between the continent and each island, the column means of the matrix X are given by

$$w^T X = w^T V_1 B_{11} V_1^T$$

since the vector of weights w is the leading vector of the matrix U and the inner product with the columns of V is zero except for the first, which is the scalar $w^T V_1 = \sum_k w_k V_{1k} = 1$. Since V_1^T is a row vector of 1's, all columns of X have the same mean value which is

$$\bar{X}_i = \bar{X} = B_{11}$$

$$= \left(\frac{(1-s)^2}{1-(1-s)^2}\right)\left(\frac{\pi(1-\pi) - X_0}{2N}\right)$$

$$\cong \frac{\pi(1-\pi) - X_0}{4Ns}$$

Maruyama (1977:136) derives this formula for a simple symmetric migration pattern. Since the column means are identical for this standard model, the statistic "distance from centroid" has expected value

$$E(R_{ii}) = \frac{X_{ii} - \bar{X}}{\pi(1-\pi) - \bar{X}} \tag{12b}$$

Combining equation (11) with equation (12) yields

$$E(R_{ii}) = \frac{H_\pi - \bar{X}}{(H_\pi - \bar{X})} - \frac{E(H_i)}{(H_\pi - \bar{X})} \tag{13}$$

Now let $H\hat{p}$ be the heterozygosity in a random mating population with gene frequencies $\hat{p}$, the island mean; then

$$H_{\hat{p}} = H_\pi - \bar{X}$$

With this substitution and with the understanding that this is an equation in expected values, (13) becomes

$$R_{ii} = \frac{H_{\hat{p}} - H_i}{H_{\hat{p}}}$$

or

$$H_i = H_{\hat{p}} - H_{\hat{p}} R_{ii} \tag{14}$$

so that distance from the sample gene frequency centroid should
be linearly related to heterozygosity with intercept $H_{\hat{p}}$ and
slope $H_{\hat{p}}$.

Example

　　We can use these formulae to calculate the expected overall
measure of differentiation within a group of subpopulations,
R_O which is equivalent to Wright's F_{st} or the G_{st} of Nei (1975).
We take the definition of R_O to be the proportional reduction
in heterozygosity due to population structure within the group
of subpopulations, that is

$$H_O = \sum_k w_k H_k = H_{\hat{p}} \, (1-R_O)$$

Now the B_{kk} above are eigenvalues of the weighted covariance matrix
$w^{\frac{1}{2}}Xw^{\frac{1}{2}}$, and the trace of this matrix is $\sum_k w_k X_{kk} = X_O$; this quantity
is given by the sum of the eigenvalues B_{kk}. We know further that
the average value of X is the leading eigenvalue $B_{11} = \bar{X}$. Now
consider the quantity describing local differentiation

$$Y = \frac{X_O - \bar{X}}{\pi(1-\pi)-X_O} = \frac{\sum_{k=2}^{t} B_{kk}}{\pi(1-\pi)-X_O} = \frac{1}{2N} \sum_{k=2}^{t} \lambda_k^2 \Big/ 1-\lambda_k^2$$

Using the facts that

$$E(H_O) = H_\pi - X_O$$

$$E(H_{\hat{p}}) = H_\pi - \bar{X}$$

this statistic is estimated by

$$Y = \frac{H_{\hat{p}} - H_O}{H_O} = \frac{R_O}{1-R_O}$$

so that

$$R_O = Y \Big/ 1 + Y$$

　　For example a collection of t islands each with endogamy
(1-m) and each receiving a fraction m/(t-1) of its members
from each other island has the migration matrix

$$1-m \quad m/_{t-1} \quad m/_{t-1} \; \cdots$$

$$m/_{t-1} \quad 1-m$$
$$\vdots$$

and eigenvalues $\lambda_1^* = 1$, $\lambda_j^* = 1 - {tm}/{t-1}$ $j = 2,3 \ldots t$
(Carmelli and Cavalli-Sforza 1976. Our assumption above of
distinct eigenvalues facilitates a perturbation later, but it
is unnecessary here or in anything so far). Hence, assuming
the immigration s is much smaller than m/(t-1) and that terms
in m^2 are small enough to be ignored,

$$Y \cong \frac{t-1}{2N} \frac{(t-1-2tm)}{2tm}$$

so that

$$R_o = \frac{Y}{1+Y} = \frac{(t-1)(t-1-2tm)}{4Nmt + (t-1)(t-1-2m)}$$

When there are a large number of islands (t>>1) and m is
small,

$$R_o \cong \frac{1}{4nm + 1} \qquad n = \frac{N}{t}$$

which is a well known formula. However if there are, say five
islands and the endogamy is 80% so that m = .2, then

$$R_o = \frac{4(4-2)}{4N + 4(4-2)} = \frac{1}{.5N + 1}$$

rather than

$$R_o \cong \frac{1}{.16N + 1}$$

as given by the usual approximation. For n = 50 the exact and
approximate figures are $R_o \sim \frac{1}{26}$ and $R_o \sim \frac{1}{9}$, that is the familiar
formula overestimates the expected amount of local differentiatio
by a factor of three. If immigration from outside the group of
islands is large it cannot be ignored as we have, and the effect
on the eigenvalues must be incorporated in the computation, so
that

$$\lambda_j = \lambda_j^* - s = 1 - s - \frac{tm}{t-1} \qquad \text{for} \quad j \neq 1.$$

Heterozygosity and Distance in Modified Models

We now examine the validity of the regression of heterozygosity on distance under two modifications of the assumptions
of the standard model; first we discuss the case when the exchang
with the continent varies from island to island, and we then
discuss the case where the exchange s is the same for each island
but the gene frequencies of the constant source vary from island
to island.

Variable Exchange with the Continent

We assume here that the entries of the diagonal matrix S are not constant, and we write $S = \bar{S} + \delta S$ where the constant entries of $\bar{S}$ are the average of the exchange rates and those of (δS) are small deviations with zero mean. As before

$$M = M* - \bar{S}$$

and we study the equilibrium

$$X = (M + \delta S)^T (X + D)(M + \delta S)$$

by regarding the matrix δS as a perturbation of the matrix M, deriving the first order approximation to the spectral decomposition of $(M + \delta S)$, showing that a form like (10) still holds, and finally showing that heterozygosity and distance from centroid are not simply related in this case.

The first order approximations to the left eigenvectors V' of $(M + \delta S)^T$ are *(Pease 1965, Bellman 1970)*

$$V'_i = V_i + \sum_{j \neq i} a_{ij} V_j$$

$$a_{ij} = \frac{V_j^T W(\delta S) V_i}{\lambda_i - \lambda_j} = -a_{ji}$$

The corresponding approximations to the right eigenvectors U' are

$$U'_i = U_i + \sum_{j \neq i} a_{ij} U_j$$

The approximations to the eigenvalues are

$$\lambda'_j = \lambda_j + V_j^T W(\delta S) V_j$$

At equilibrium the genetic covariance matrix satisfies

$$X = M'^T (X + D) M'$$

$$\underset{\sim}{X} = \underset{\sim}{M}'^T \underset{\sim}{X} + \underset{\sim}{M}'^T \underset{\sim}{D} \qquad (15)$$

to first order, where the prime indicates the perturbation approximation to $(M + \delta S)$ and $\underset{\sim}{X}$ is the Kronecker form of X as in equation (7).

The solution to (15) can now be written

$$X_{ij} = \sum_{k,l} \frac{\lambda'_k \lambda'_l}{1 - \lambda'_k \lambda'_l} V'_{il} V'_{jk} \{\sum_m U'_{ml} U'_{mk} D_{mm}\}$$

As before, the sum in { } on the right is crucial to the simplification of this form. It is

$$\frac{\pi(1-\pi)-X_O}{2N} \sum_m U'_{ml} U'_{mk} w_m^{-1}$$

$$= \frac{\pi(1-\pi)-X_O}{2N} \sum_m [U_{ml} U_{mk} w_m^{-1} + w_m^{-1} \sum_{f \neq l} U_{mk} U_{mf} a_{lf}$$

$$+ w_m^{-1} \sum_{g \neq k} U_{ml} U_{mg} a_{kg} + O(a^2)]$$

There are two cases to consider, the case when k=l and the case when k≠l. Recalling that $\sum_m U_{mx} U_{my} w_m^{-1}$ is equal to one when x = y and 0 otherwise we have

Case 1: k = l

$$\frac{\pi(1-\pi)-X_O}{2N} \sum_m [U_{mk}^2 w_m^{-1} + w_m^{-1} \sum_{f \neq k} U_{mk} U_{mf} a_{kf}$$

$$+ w_m^{-1} \sum_{g \neq k} U_{mk} U_{mg} a_{kg} + O(a^2)]$$

$$= \frac{\pi(1-\pi)-X_O}{2N} [1 + 0 + 0 + O(a^2)]$$

$$= \frac{\pi(1-\pi)-X_O}{2N} \text{ to first order}$$

Case 2: k ≠ l

$$= \frac{\pi(1-\pi)-X_O}{2N} \sum_m [U_{ml} U_{mk} w_m^{-1} + w_m^{-1} \sum_{f \neq l} U_{mk} U_{mf} a_{lf}$$

$$+ w_m^{-1} \sum_{g \neq k} U_{ml} U_{mg} a_{kg} + O(a^2)]$$

$$= \frac{\pi(1-\pi)-X_O}{2N} \sum_m [0 + w_m^{-1} U_{mk}^2 a_{lk} + w_m^{-1} U_{ml}^2 a_{kl} + O(a^2)]$$

$$= 0 \text{ to first order, since } a_{lk} = -a_{kl}$$

Thus this sum is still, to first order, the constant $\dfrac{\pi(1-\pi)-X_0}{2N}$
multiplied by the k,ℓ*'th entry of an identity matrix, and the*
representation

$$X = V'B'V'^{T} \tag{16}$$

is valid, with B' *a diagonal matrix*

$$B'_{kk} = [\frac{\lambda'^2_k}{1-\lambda'^2_k}]\ [\frac{\pi(1-\pi)-X_0}{2N}]$$

We will use this form to calculate the effect of the variation
in s on the expected heterozygosity of the i'th island, pro-
portional to X_{ii}*, and on the distance from centroid of the i'th*
island, proportional to $R_{ii} \propto X_{ii} - 2\bar{X}_i + \bar{X}$.

From (16)

$$X_{ii} = \sum_k V'^2_{ik} B'_k$$

$$= \sum_k B'_k\ (V_{ik} + \sum_{l \neq k} a_{kl}V_{il})^2$$

$$= \sum_k B'_k V^2_{ik} + \sum_{k=2}^{t} \sum_{\ell=k+1}^{t}\ (B'_k - B'_\ell)(2a_{k\ell}V_{ik}V_{i\ell}) + O(a^2) \tag{17}$$

since $a_{k\ell} = -a_{\ell k}$. *It will be convenient to separate here the*
effects of the perturbations to and from the first dimension:

$$X_{ii} = \sum_k B'_k V^2_{ik} + \sum_{\ell=2}^{t} (B'_1 - B'_\ell)(2a_{1\ell}V_{i1}V_{i\ell})$$

$$+ \sum_{\substack{k=2 \\ \ell=k+1}}^{t} (B'_k - B'_\ell)(2a_{k\ell}V_{ik}V_{i\ell}) \tag{17a}$$

In order to compute the distance R_{ii} *we need the column*
average

$$\bar{X}_i = \sum_j w_j \sum_k B'_k V'_{ik} V'_{jk}$$

Recall that the weights w are the entries of the first
column of U, so the inner product

$$\sum_j w_j V_{jk} = \begin{cases} 1 & \text{if } k = 1 \\ 0 & \text{if } k \neq 1 \end{cases}$$

so

$$\bar{X}_i = \sum_j w_j \sum_k B'_k (V_{ik} + \sum_{x \neq k} a_{kx} V_{ix})(V_{jk} + \sum_{x \neq k} a_{kx} V_{jx})$$

$$= B'_1 (V_{i1} + \sum_{x \neq 1} a_{1x} V_{ix}) + \sum_{y \neq 1} B'_y (V_{iy} + \sum_{x \neq y} a_{yx} V_{ix}) a_{y1}$$

$$= B'_1 + \sum_{k \neq 1} (B'_1 - B'_k)(a_{1k} V_{ik}) + O(a^2) \qquad (18)$$

and

$$\bar{X} = \sum_i w_i \bar{X}_i = B'_1 \qquad (19)$$

  Combining equations (17), (18), *and* (19) *recalling that* $V_{i1}=1$ *for all i, we have*

$$R_{ii} \propto X_{ii} - 2\bar{X}_i + \bar{X}$$

$$= \sum_{k=2}^{t} B'_k V_{ik}^2 + \sum_{k=2}^{t} \sum_{\ell=k+1}^{t} (B'_k - B'_\ell)(2a_{k\ell} V_{ik} V_{i\ell}) \qquad (20)$$

*The effect of the variation in s appears to these equations in
two ways; the perturbation to the eigenvalues B; this is very
small in numerical experiments; and in the perturbations to the
eigenvectors. Heuristically equation 17a says, neglecting the
perturbation to the eigenvalues, that X_{ii} and consequently hetero
zygosity are changed by the perturbations due to the first
dimension plus those due to dimensions two and higher. Equation
(20) says that the distance from centroid is only changed by the
perturbations among the second and higher dimensions. We have no
explicit demonstration, but regularly in numerical calculations
these latter terms are very small. The major effect of the
variation in s is to change the expected heterozygosity while
the change in the genetic distances are inconsequential. This
means that the use of genetic distance for inference about
migration and gene flow is not seriously affected by variation
among islands or subpopulations in exchange with an outside
constant source. This conclusion, however, depends upon the use
of the kind of genetic distance we are using here, i.e., one
related to chi-square kinds of statistics. The kind of genetic
distance developed by Nei (1975) has many advantages for interspecific investigations, but it is more closely related to
heterozygosity statistics than to genetic distance as we are
using the term, and we expect it to be severely distorted by
uneven systematic pressure.*

  *On the other hand, an examination of a plot of heterozygosity
versus distance from the centroid is expected to provide useful
information about non-uniform exchange, because islands with high*

or low exchange should be outliers from the regression line; those with greater than average contact will be more heterozygous than the prediction from genetic distance, while isolated islands will show low heterozygosity.

Uniform Exchange with Several Continents

In this case π is a column vector of gene frequencies ; π_i is the frequency of the allele in immigrants to the i'th island. Each island exchanges the same fraction s of its population with its continent. If $\bar{p}_i$ is the deterministic mean allele frequency of the island, the vector $\bar{p}$ solves, at equilibrium.

$$\bar{p} = M^T\bar{p} + s\pi$$

$$\bar{p} = (I-M^T)^{-1}s\pi \tag{21}$$

and the overall mean is

$$\sum_k w_k\bar{p}_k$$

$$= \sum_k w_k \ [1/(1-\lambda_k)] \ U_k V_k^T s\pi$$

$$= \bar{\pi} \ [definition]$$

The deterministic dispersion around this immigrant mean $\bar{\pi}$ *is*

$$(\bar{p}-\bar{\pi})(\bar{p}-\bar{\pi})^T$$

$$\sum_{k=1}^{t} \ V_k \ [1/(1-\lambda_k)] \ U_k^T \ s\pi\pi^T sU \ [1/(1-\lambda_k)] \ V_k^T$$

$$= X_{mig} \ [definition]$$

This matrix has the property that any row or column average is zero; the column averages are given by the product $w_k^T \ X_{mig}$, *but the leading element of* X_{mig} *is the matrix of column vectors* V_k *for k=2, 3, ..., t;* w_k *is the first row of* U^T *and is orthogonal to all these columns of* V *so the product is a row of zeroes.*

If reproduction and genetic drift alternate with migration as before, then the dispersion around $\bar{\pi}$ will increase. Let Δ_k be the deviation of p_k from its deterministic value $\bar{p}_k$. Then

$$p_k(t + 1) = \bar{p}_k + \Delta_k(t + 1) = M^T \ [\bar{p}_k + \Delta_k \ (t)] + s\pi + \varepsilon_k \ (t)$$

and, using (21)

$$\Delta_k(t + 1) = M^T \Delta_k(t) + \varepsilon_k(t)$$

The total genetic covariance matrix around the population mean
$\bar{\pi}$ *is the expectation of*

$$(\bar{p} - \bar{\pi} + M^T\Delta + \varepsilon)(\bar{p} - \bar{\pi} + M^T\Delta + \varepsilon)^T$$

$$= X_{mig} + X_{drift} = X \tag{22}$$

where X_{drift} *is the equilibrium solution of equation* (7) *as
before, except that matrix D of new variance will use the
average diagonal entry of the sum of* X_{mig} *and* X_{drift} *for* X_0,
rather than just the average diagonal value of X_{drift} *as in
equation* (7); *the additional dispersion due to the heterogeneity
of continental input slows down very slightly the rate of loss
of heterozygosity.*

*Under this model the row average of X is the same for each
row and is equal to the overall mean* B_{11} *as in the standard
model, and the regression of heterozygosity on distance from
centroid is, as in* (14)

$$H_i = H_{\hat{p}} - H_{\hat{p}} R_{ii}$$

but the average heterozygosity of islands H_i *will be less and
the genetic distances* R_{ii} *will be greater than in the standard
model by an amount depending on the magnitude of* X_{mig}. *If
adequate information were available, equation* (22) *might be
used to untangle the effects of population structure and of
immigration on gene frequency variation in a region. For
example in this case the principal components of the weighted
sample covariance matrix* $W^{\frac{1}{2}}RW^{\frac{1}{2}}$ *will no longer be the eigenvectors
2, 3, ..., t of the migration matrix; these vectors V of M will
be eigenvectors of that part of R due to* X_{drift}, *and there is
perhaps some possibility of using this fact to separate these
influences.*

LITERATURE CITED

BELLMAN, R. 1970. Introduction to Matrix Analysis. McGraw-Hill
BODMER, W. and L. CAVALLI-SFORZA. 1968. A migration Matrix
    Model for the Study of Random Genetic Drift. Genetics.
    59:565-592.
CARMELLI, D., and L.L. CAVALLI-SFORZA. 1976. Some Models of
    Population Structure and Evolution. Theor. Pop. Biol.
    9(3):329-359.

COURGEAU, D. 1974. Migration, a chapter in A. Jacquard, Genetic Structure of Populations. Springer.

ENDLER, J., 1977. Geographic Variation, Speciation, and Clines. Princeton University Press, Princeton, N.J.

FITCH, W. and J.V. NEEL. 1969. Phylogenetic Relationships of some Indian tribes of Central and South America. American Journal of Human Genetics. 21:384-394.

GERSHOWITZ, H., M. LAYRISSE, Z. LAYRISSE, J. NEEL, N. CHAGNON, M. AYRES. 1972. The Genetic Structure of a Tribal Population, the Yanomamo Indians. II. Annals of Human Genetics. 35:261-269.

HAMILTON, W.D. 1964. The Genetical Theory of Social Behavior. I, II. Journal of Theoretical Biology. 7:1-52.

HARPENDING, H. and W. CHASKO. 1976. Heterozygosity and Population Structure in southern Africa. *In:* The Measures of Man, E. Giles and J. Friedlander, eds., pp. 214-229. Peabody Museum Press, Cambridge.

HARPENDING, H. and T. JENKINS. 1973. Genetic Distances Among Southern African Populations. *In:* Methods and Theories of Anthropological Genetics, M. Crawford and P. Workman, eds., pp. 177-199. University of New Mexico Press.

HARPENDING, H. and T. JENKINS. 1974. !Kung Population Structure. *In:* Genetic Distance, J. Crow and C. Denniston, eds. Plenum.

JORDE, LYNN B. 1979. The Genetic Structure of the Aland Islands, Finland. Ph.D. Dissertation, University of New Mexico.

KEMENY, J. and J. SNELL. 1960. Finite Markov Chains. Van Nostrand.

KIMURA, M. 1968. Evolutionary Rate at the Molecular Level. Nature. 217:624.

KING, J. and T. JUKES. 1969. Non-Darwinian Evolution. Science. 164:788.

MALECOT, G. 1969. The Mathematics of Heredity. Freeman.

MARUYAMA, T. 1977. Stochastic Problems in Population Genetics. Springer.

NEI, N. 1975. Molecular Population Genetics and Evolution. North-Holland.

PEASE, J. 1965. Methods of Matrix Algebra. Academic Press.

SLATKIN, M. 1980. The Distribution of Mutant Alleles in a Subdivided Population. Genetics. 95:503-524.

SMITH, C.A.B. 1969. Local Fluctuations in Gene Frequencies. Annals of Human Genetics. 32:251-260.

STANLEY, S. 1979. Macroevolution. Freeman

TANIS, R.J., J.V. NEEL, H. DOREY and M. MORROW. 1973. The Genetic Structure of a Tribal Population, the Yanomamo Indians. American Journal of Human Genetics, pp. 655-676.

WARD, R., 1972. The Genetic Structure of a Tribal Population, the Yanomamo Indians, V. Annals of Human Genetics. 36:21-43.

WARD, R.H., HENRY GERSHOWITZ, MIGUEL LAYRISSE and JAMES V. NEEL. 1975. The Genetic Structure of a Tribal Population, the Yanomamo Indians, XI. American Journal of Human Genetics. 27:1-30.

WEISS, K. and T. MARUYAMA. 1976. Archaeology, Population
    Genetics and Studies of Human Racial Ancestry. American
    Journal of Physical Anthropology. 44(1):31-50.
WEITKAMP, L.R., T. ARENDS, M.L. GALLANGO, J.V. NEEL, J. SCHULTZ
    and D.C. SHREFFLER. 1972. The Genetic Structure of a Tribal
    Population, the Yanomamo Indians, III. Annals of Human
    Genetics. 35:271-279.
WEITKAMP, L.R. and J.V. NEEL. 1972. The Genetic Structure of
    a Tribal Population, the Yanomamo Indians, IV. Annals of
    Human Genetics. 35:433-444.
WORKMAN, P., H. HARPENDING, J. LaLOUEL, C. LYNCH, J. NISWANDER
    and R. SINGLETON. 1973. Population Studies on Southwestern
    Indian Tribes, VI. In: Genetic Structure of Populations,
    N.E. Morton, ed., pp. 166-194. University of Hawaii Press,
    Honolulu.
WRIGHT, S. 1931. Evolution in Mendelian Populations. Genetics,
    16:97-159.
YOKOYAMA, S. 1980. The Effect of Social Selection on Population
    Dynamics of Rare Deleterious Genes. Heredity. 45:271-280.

OF CLOCKS AND CLADES, OR
A STORY OF OLD TOLD BY GENES OF NOW

Francisco J. Ayala

Department of Genetics
University of California
Davis, California

The dramatic discoveries of molecular biology over the last three decades have provided a new understanding of evolutionary processes. The reconstruction of phylogeny (evolutionary history) and the timing of important evolutionary events have become possible with much greater precision than ever before.

Informational macromolecules (DNA, RNA, and proteins) contain, in the sequence of their component units, considerable evolutionary information. Events that occurred in a more or less remote past, and for which fossil remains may be lacking, are recorded in the nucleotide sequence of genes and in the amino acid sequence of proteins. Comparative study of genes and proteins has, therefore, made possible an accurate reconstruction of evolutionary events. And the events are being timed by a fascinating new development: the "molecular clock" of evolutionary processes.

INFORMATION AND CONFORMATION

In 1909 Wilhelm Johannsen introduced the distinction between phenotype and genotype. The phenotype of an organism is its appearance -- what we can observe: its morphology, physiology and behavior. The genotype is the genetic constitution it has inherited (Ayala and Kiger, 1980). During the lifetime of an individual, the phenotype changes; the genotype,

however, remains constant. Moreover, a given genotype does
not unambiguously specify a given phenotype or a given
sequence of phenotypes as the life of the individual proceeds.
This is because the phenotype is the result of development
(ontogeny) during which complex networks of interactions
between the genotype and the environment take place. What
the genotype determines is the range (or "norm") of reaction,
the range of phenotypes that *may* develop. Which phenotype,
or sequence of phenotypes, will be realized depends on the
environment in which development takes place. Development
in the broadest sense includes, of course, the complete life
of the individual.

Evolution is the process of change of organisms through
the generations. The evolutionary process involves the gradual
change in both genotypes and phenotypes. Most apparently,
evolution is change in the phenotype of organisms through the
generations. That which lives -- what is born, interacts with
the environment and with other organisms, and reproduces --
is the phenotype, the organism itself. But phenotypic change
is of no evolutionary consequence unless it is underlain by
genotypic change, because only the genotype is transmitted to
the following generations. "Acquired" characteristics are
not inherited. Thus, in a very fundamental sense, evolution
is genotypic change and the history of evolution is determined
by the sequence of genotypes that have existed through time.

The contrast between genotype and phenotype may also be
formulated as a dichotomy between information and conformation
(see Kendrew, 1968). The genotype is the information, the
set of instructions, passed from one generation to the next.
The phenotype is the conformation or configuration that
realizes the genetic blueprint in a given set of environmental circumstances. Formulating the dichotomy at hand in

terms of information versus conformation tips off a very
important property of the genotype: not only does it convey
ontogenetic information, but in addition it stores evolutionary information. The genotype contains the instructions that
guide the development of the individual but it also contains
a record of its evolutionary history. To be sure, the conformation of organisms does also reflect their past history,
but this reflection is muddled by the interactions between
the genotype and the environment, and it cannot be quantified
in a precise manner. On the contrary, the information contained in the genotype is not affected by developmental interactions, it is readily quantifiable, and, as we shall see,
makes it possible to compare even the most diverse organisms.

THE RECONSTRUCTION OF PHYLOGENY

The study of fossil remains of organisms living in the
past provides definite clues of the phylogeny (*i.e.*,
evolutionary history) of a group of organisms, but the fossil
record is always incomplete and often very limited or
altogether lacking. Phylogenetic inferences may also be made
from the comparative study of living organisms (Simpson, 1953).
The logical basis of these inferences is simple. Because
evolution is by-and-large a process of divergence, organisms
sharing a recent common ancestor are likely to be more similar
to each other than organisms with a common ancestor only in
a more remote past: relative degrees of similarity are used
to infer recency of common descent. Assume, for example,
that we compare three species and find that two of them, A
and B, are much more similar to each other than they are to
C; we would infer that the lineage leading to C separated
from a lineage going to both A and B, before this lineage

split again into two. Needless to say, the reconstruction of
evolutionary history from the study of living organisms is a
task far from simple: rates of evolutionary change may be
different at different times, or in different groups of
organisms, or with respect to different features of the same
organisms. Moreover, resemblances due to common descent must
be set apart from resemblances due to similar ways of life,
to life in the same or similar habitat, or to accidental
convergence.

Until about two decades ago, the biological discipline
that provided most information about evolutionary history
was comparative anatomy, but additional knowledge was
obtained from embryology, cytology, ethology, and biogeography. These biological disciplines all study the conformation
of organisms at a level far removed from the genotype. The
advances of molecular biology have made possible, in recent
years, the comparative study of proteins and nucleic acids --
DNA and RNA (Dobzhansky *et al.*, 1977). The DNA is the re-
pository of hereditary (evolutionary and developmental)
information. Proteins are part of the conformation of the
organism, but their relationship to the DNA is so immediate
that they closely reflect the hereditary information. This
reflection is not perfect because the genetic code is
redundant and, hence, some differences in the DNA do not
yield differences in the proteins. Moreover, it is not
complete because, as we now know, a large fraction (about 90
percent) of the DNA does not code for proteins. Nevertheless,
because proteins are so closely related to the information
contained in the DNA, they, as well as the nucleic acids,
are called "informational macromolecules." Needless to say,
all relevant information should be used in the reconstruction
of phylogeny, but informational macromolecules are a much more

powerful tool for the study of evolutionary history than
comparative anatomy and the other disciplines that study the
conformation of organisms at a level far removed from the
hereditary information.

Informational macromolecules retain indeed a considerable
amount of evolutionary information. Nucleic acids and proteins
are linear molecules made up of units, nucleotides in the case
of nucleic acids, amino acids in the case of proteins. The
sequence of the units contains information in a similar way as
the sequence of letters and punctuation marks in a paragraph
contains information. Comparison of two macromolecules estab-
lishes the number of units which are different in them.
Because evolution usually occurs by changing one unit at a
time, the number of differences is an indication of the
recency of common ancestry. Changes in evolutionary rates
may create difficulties, but macromolecular studies have two
notable advantages over comparative anatomy and the other
classical disciplines. One is that the information is more
readily quantifiable: the number of units that are different
is readily established when the sequence of units is known
for a given macromolecule in different organisms. The other
advantage is that even very different sorts of organisms can
be compared. There is very little that comparative anatomy
can say when organisms as diverse as yeasts, pine trees, and
human beings are compared; but there are homologous macromolecules that can be compared in all three.

Informational macromolecules provide information not only
about the topology of evolutionary history -- *i.e.*, about
the splitting of lineages, or *cladogenesis* -- but also about
the amount of genetic change that has occurred in any given
evolution lineage or *anagenesis*. It might seem at first that
the genetic study of anagenesis is impossible, because it
would require the comparison of organisms that lived in the

past with living organisms. Organisms of the past are sometimes preserved as fossils, but their DNA and proteins are
largely disintegrated. Nevertheless, comparisons between
living species provide information about anagenesis. Consider
two contemporary species, *C* and *D*, evolved from a common
ancestral species, *B*. Assume that we find that *C* and *D*
differ by *x* amino acid subsitutions in a certain protein, say
myoglobin. It is reasonable to assume, as a first approximation, that *x*/2 substitutions have taken place in each of the
two evolutionary lineages, *i.e.*, from *B* to *C* and from *B* to
D (fig. 1).

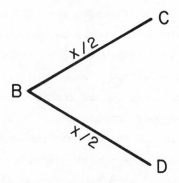

FIGURE 1.   *Inference of anagenetic evolution from*
cladogenetic data. C and D are two
contemporary species having B as a common
ancestral species. If the amount of genetic
differentiation between C and D is x, *we can*
assume, as a first approximation, that half
of the change occurred in each of the two
lineages.

The assumption that equal amounts of change have occurred
in the two lineages can be removed. Suppose that a third
contemporary species *E*, is compared with *C* and *D* and that the
number of amino acid differences between the myoglobin
molecules of the three species are as follows:

C and D =   4

C and E = 11

D and E =   9

If the phylogeny of the three species is as shown in fig. 2, we can estimate the number of substitutions that have occurred in each of its branches. Let us use x and y to denote the number of amino acid differences between B and C and between B and $D,$ respectively and z to denote the number of differences between A and B *plus* those between A and E. We have, then, the following three equations:

$$x + y =   4$$

$$x + z = 11$$

$$y + z =   9$$

Solving these equations, we obtain $x = 3,$ $y = 1,$ and $z = 8.$

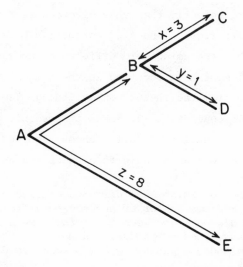

FIGURE 2.   *Estimated amounts of anagenetic change in the phylogeny of three contemporary species.*

The procedure becomes more complicated when many more contemporary species are involved, but the conceptual basis

for estimating anagenetic change is the same. As a concrete
example, consider cytochrome *c*, a protein involved in cell
respiration. The sequence of amino acids is known in many
organisms, from bacteria and yeast to insects and humans. In
animals, cytochrome *c* consists of 104 amino acids. When the
amino acid sequences of humans and rhesus monkeys are compared,
they are found to be different at position 66 (isoleucine in
humans, threonine in rhesus monkeys), but identical at the
other 103 positions. When horses are compared with humans 12
amino acid differences are found, but when horses are compared
with rhesus monkeys there are only 11 amino acid differences
(table 1, above the diagonal).

Even if we did not know anything else about the
evolutionary history of mammals, we would conclude that the
lineages of humans and rhesus monkeys diverged from each
other much more recently than they diverged from the horse
lineage (compare figs. 3 and 4). Moreover, it is possible
to conclude that the amino acid difference between humans and
rhesus monkeys must have occurred in the human lineage after
its separation from the rhesus-monkey lineage (see fig. 3).

The genetic code (table 2) makes it possible to calculate the
minimum number of nucleotide differences required to change
from a codon for one amino acid to a codon for another. At

*Table 1.   Number of amino acid differences (above the diagonal)
and minimum number of nucleotide differences (below
the diagonal) between the cytochrome c molecules of
humans, rhesus monkeys, and horses.   The cytochrome
c in these organisms has 104 amino acids.*

	Human	Rhesus monkey	Horse
Human	--	1	12
Rhesus monkey	1	--	11
Horse	15	14	--

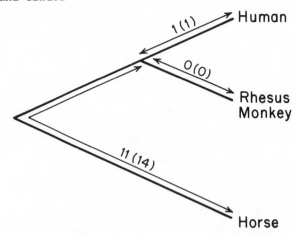

FIGURE 3. Anagenetic change in the evolution of cytochrome
          c from humans, rhesus monkeys, and horses. The
          numbers indicate the amino acid substitutions
          (and, in parentheses, the minimum number of
          nucleotide substitutions) that have taken place
          in each branch of the phylogeny.

position 19 of cytochrome *c*, humans and rhesus monkeys have
isoleucine, but horses have valine. Isoleucine may be encoded
by any one of the three codons AUU, AUC, and AUA, and valine,
by any one of the four codons GUU, GUC, GUA, and GUG. Thus,
one single nucleotide-pair substitution (from A to G in the
first position) is sufficient to change a codon for isoleucine
to a codon for valine. At position 20, humans and rhesus
monkeys have methionine (AUG), while horses have glutamine
(CAA or CAG); therefore, at least two nucleotide-pair
substitutions (in the first and second positions) must have
occurred to change the methionine codon to a codon for
glutamine. The minimum numbers of nucleotide-pair substitutions
required to account for the amino acid differences between the
cytochrome *c* molecules of humans, rhesus monkeys, and horses
are shown (below the diagonal) in table 1 and (in parentheses)
in fig. 3.

The methods developed for estimating genetic change during

Francisco J. Ayala

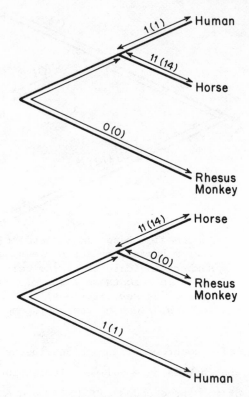

FIGURE 4.   *Two theoretically possible phylogenies of*
humans, rhesus monkeys, and horses. The numbers
of amino acid (and nucleotide) substitutions
required in each branch to account for the
cytochrome c *sequences indicate that neither of*
these two phylogenies is likely to be correct.

evolution from the study of informational macromolecules
include DNA hybridization, protein sequencing, immunology,
and electrophoresis (Dobzhansky *et al.*, 1977). In addition,
during the last five years, methods have been developed that
allow ascertaining the nucleotide sequence of individual genes
and of other DNA segments (Sanger *et al.*, 1978; Maniatis *et*
al., 1978). The study of the nucleotide sequences of genes
has apported fascinating, and indeed unexpected, insights.
The following pages review these various methods for reconstructing phylogeny and for estimating the amount of genetic

Table 2. *The genetic code gives the correspondence between the 64 possible codons in messenger RNA and the amino acids (or termination signals). The nitrogen base thymine does not exist in RNA, where uracil (U) takes its place; the other three nitrogen bases in messenger RNA are the same as in DNA: adenine (A), cytosine (C), and guanine (G). The 20 amino acids making up proteins are as follows: alanine (Ala), arginine (Arg), asparagine (Asn), aspartic acid (Asp), cystein (Cys), glycine (Gly), glutamic acid (Glu), glutamine (Gln), histidine (His), isoleucine (Ile), leucine (Leu), lysine (Lys), methionine (Met), phenylalanine (Phe), proline (Pro), serine (Ser), threonine (Thr), tyrosine (Tyr), tryptophane (Trp), and valine (Val).*

SECOND LETTER

		U	C	A	G	
FIRST LETTER	**U**	UUU UUC } Phe UUA UUG } Leu	UCU UCC UCA UCG } Ser	UAU UAC } Tyr UAA UAG } Stop	UGU UGC } Cys UGA Stop UGG Trp	U C A G
	C	CUU CUC CUA CUG } Leu	CCU CCC CCA CCG } Pro	CAU CAC } His CAA CAG } Gln	CGU CGC CGA CGG } Arg	U C A G
	A	AUU AUC AUA } Ile AUG Met	ACU ACC ACA ACG } Thr	AAU AAC } Asn AAA AAG } Lys	AGU AGC } Ser AGA AGG } Arg	U C A G
	G	GUU GUC GUA GUG } Val	GCU GCC GCA GCG } Ala	GAU GAC } Asp GAA GAG } Glu	GGU GGC GGA GGG } Gly	U C A G

THIRD LETTER

change that has occurred in each lineage.

DNA HYBRIDIZATION

A technique that estimates the overall similarity between
the DNA of various organisms is DNA "hybridization" (Hoyer *et al.*,

1964). DNA radioactively labelled, which has been "melted" (*i.e.*, dissociated) and fractioned can be reacted with various amounts of melted DNA from a different species. Homologous sequences will hybridize to form duplexes; the extent of the reaction gives an estimate of the proportion of DNA sequences that are homologous (fig. 5).

Sequences forming duplexes need not be complementary for every nucleotide. The proportion of noncomplementary nucleotides in interspecific DNA duplexes can be estimated by the rate at which the DNA strands sepa.ate at increasing temperatures (Laird and McCarthy, 1968). The critical parameter, called *thermal stability* (T_S), is the temperature at which 50 percent of the duplex DNA has dissociated (fig. 6). The difference (ΔT_S) between the T_S of hybrid and control DNA is approximately directly proportional to the proportion of unpaired nucleotides in the hybrid DNA, so that $1°C$ $\Delta T_S \approx 1$ percent mismatched nucleotides (Laird *et al.*, 1969; McCarthy and Farquhar (1974). The results of comparing the DNA of various primates first with human DNA and then with green monkey DNA (Table 3) serve to estimate the percentage of nucleotide-pair substitutions that have occurred during primate evolution (fig. 7).

FROM PROTEIN SEQUENCE TO PHYLOGENY

The amino acid sequence of the cytochromes *c* of many organisms are known. Phylogenies can, then, be constructed based on the number of amino acid differences between species. However, the amino acid sequence of a protein contains more information than is reflected in the number of amino acid differences. This is because, as pointed out above, the replacement of one amino acid by another requires in some

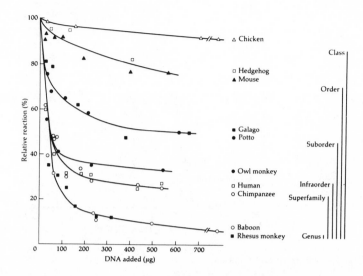

FIGURE 5.   *Homology between DNA sequences of rhesus monkeys and those of various other species. The amount of filter-bound DNA from rhesus monkey was 130 μg. The DNA in solution included half a microgram of labeled rhesus DNA and varying amounts of DNA from other species. The proportions of nonhomologous DNA are indicated by the bars at right, where the taxonomic separation between rhesus monkeys and the other species is indicated. (After Kohne et al., 1972).*

Table 3.   *Percent nucleotide differences between the DNA of various primates and the DNA of humans and green monkeys. (Data from Kohne et al., 1972.)*

| Species tested | Tester DNA from: | |
	Human	*Green monkey*
Human	0	9.6
Chimpanzee	2.4	9.6
Gibbon	5.3	9.6
Green monkey	9.5	0
Rhesus monkey	--	3.5
Capuchin	15.8	16.5
Galago	42.0	42.0

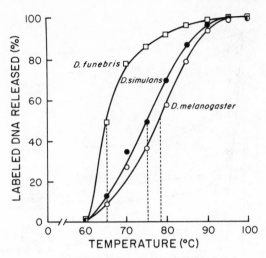

FIGURE 6.   Thermal stability profiles of DNA duplexes
            having one strand from Drosophila melanogaster
            and the other from the species indicated. The
            critical parameter called thermal stability
            (T_s) is the temperature at which 50 percent of
            the duplex DNA has dissociated. The difference
            (ΔT_s) between the T_s of hybrid DNA and that of
            nonhybrid DNA (D. melanogaster with D. melano-
            gaster) corresponds approximately to the
            percentage of mismatched nucleotide pairs in
            the hybrid DNA duplex. The T_s for the nonhybrid
            duplex DNA is 78° C, for the D. melanogaster-
            D. simulans duplex DNA 75° C, and for the
            D. melanogaster-D. funebris duplex DNA 65° C.
            Thus, the proportion of nucleotide pairs
            different from D. melanogaster is three percent
            for D. simulans and 13 percent for D. funebris.
            (After Laird and McCarthy, 1968).

cases no more than one nucleotide substitution in the DNA, but

in other cases it requires at least two nucleotide changes in

the corresponding triplets (codons).

    The minimum numbers of nucleotide differences necessary

to account for the amino acid differences in the cytochromes

c of twenty organisms are given in table 4 (Fitch and

Margoliash, 1967, 1970). A phylogeny based on that data

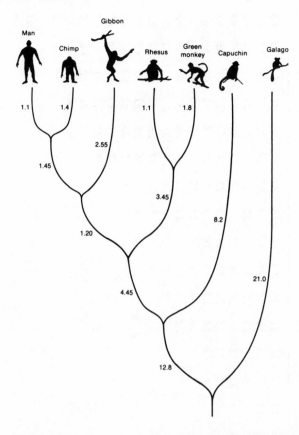

FIGURE 7.   *Phylogeny of a number of species of primates
            based on DNA hybridization techniques. The
            numbers on the branches estimate the per-
            centage of nucleotide-pair substitutions that
            have occurred during evolution. (After Kohne
            et al., 1972.)*

matrix, as well as the minimum numbers of nucleotide changes
required in each branch, are shown in fig. 8. These
differences are often fractions. It is obvious that a
nucleotide change may or may not have taken place but,
fractional nucleotide changes cannot occur. However, the
values given in fig. 8 are those satisfying best the data in
table 4.

Table 4. Minimum number of nucleotide differences in the genes coding for cytochromes c in 20 organisms. (Data from Fitch and Margoliash, 1967).

Organism	2	3	4	5	6	7	8	9	10	11	12	13	14	15	16	17	18	19	20
1. Human	1	13	17*	16	13	12	12	17	16	18	18	19	20	31	33	36	63	56	66
2. Monkey		12	16*	15	12	11	13	16	15	17	17	18	21	32	32	35	62	57	65
3. Dog			10	8	4	6	7	12	12	14	14	13	30	29	24	28	64	61	66
4. Horse				1	5	11	11	16	16	16	17	16	32	27	24	33	64	60	68
5. Donkey					4	10	12	15	15	15	16	15	31	26	25	32	64	59	67
6. Pig						6	7	13	13	13	14	13	30	25	26	31	64	59	67
7. Rabbit							7	10	8	11	11	11	25	26	23	29	62	59	67
8. Kangaroo								14	14	15	13	14	30	27	26	31	66	58	68
9. Duck									3	3	3	7	24	26	25	29	61	62	66
10. Pigeon										4	4	8	24	27	26	30	59	62	66
11. Chicken											2	8	28	26	26	31	61	62	66
12. Penguin												8	28	27	28	30	62	61	65
13. Turtle													30	27	30	33	65	64	67
14. Rattlesnake														38	40	41	61	61	69
15. Tuna															34	41	72	66	69
16. Screwworm fly																16	58	63	65
17. Moth																	59	60	61
18. Neurospora																		57	61
19. Saccharomyces																			41
20. Candida																			—

* The differences between the horse and either human or rhesus monkey given here (17 and 16) are greater than in Table 1 (15 and 14). The two additional nucleotide substitutions are required when all the organisms included in the present table are taken into account.

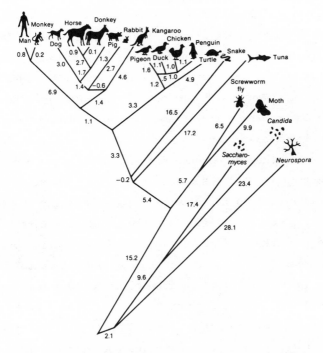

FIGURE 8.   *Phylogeny based on differences in the protein
sequence of cytochrome* c *in organisms ranging
from yeast to man.   The numbers are estimates
of the nucleotide substitutions in this
protein that have occurred during evolution.
This pattern agrees well with the pattern of
phylogenetic relationships worked out by
classical techniques of comparative morphology
and from the fossil record.   (After Fitch and
Margoliash, 1967.)*

The phylogenetic relationships shown in fig. 8 correspond
fairly well, on the whole, with the phylogeny of the organisms
as determined from the fossil record and other sources.   There
are disagreements, however.   For example, chickens appear more
closely related to penguins than to ducks and pigeons, and men
and monkeys diverge from the other mammals before the marsupial

kangaroo separates from the non-primate placentals. In spite
of these erroneous relationships, it is remarkable that the study
of a single protein yields a fairly accurate representation of
the phylogeny of 20 organisms as diverse as those in the figure.
The amino acid sequence of proteins (and the genetic information
therein contained) store considerable evolutionary information.

Cytochromes *c* are slowly evolving proteins. Organisms as
different as humans, silkworm moths, and *Neurospora* have in
common a large proportion of amino acids in their cytochrome
c. The evolutionary conservation of this cytochrome makes
possible the study of genetic differences among organisms
only remotely related. However, this same conservation makes
cytochrome *c* useless for determining evolutionary change in
closely related organisms, since these may have cytochromes *c*
that are completely or nearly identical. The primary structure
of cytochrome *c* is identical in humans and chimpanzees, which
diverged 10 to 15 million years ago; it differs by only one
amino acid replacement between humans and rhesus monkeys, whose
most recent common ancestor lived 40 to 50 million years ago.

Fortunately, different proteins evolve at different rates.
Phylogenetic relationships among closely related organisms may
be inferred by studying the primary sequences of rapidly evolving
proteins, such as fibrinopeptides in mammals (fig. 9). Carbonic
anhydrases are rapidly evolving proteins that are physiologically
important in the reversible hydration of CO_2 and in certain
secretory processes. A phylogeny based on the amino acid
sequence of carbonic anhydrase I, as well as the minimum number
of nucleotide changes in each branch, are shown in fig. 10
(Tashian *et al.*, 1976). Genetic changes in the evolution of very
closely related species can also be studied by other methods,
such as DNA hybridization, immunology, and gel electrophoresis.

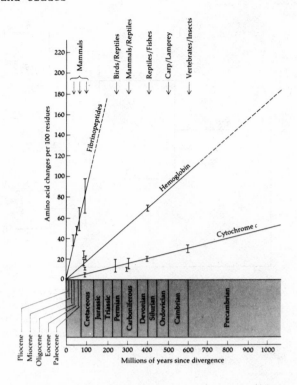

FIGURE 9.   Rates of molecular evolution of different
            proteins.   (After R.E. Dickerson, 1971).

IMMUNOLOGICAL TECHNIQUES

About one thousand sequences or partial sequences of
proteins are presently known, containing much evolutionary
information.   More sequences are obtained every year, although
the procedures for determining the primary sequences of proteins
are extremely laborious.   Other methods, such as immunological
techniques, permit estimating the degree of similarity between
proteins with much less work than amino acid sequencing
(Reichlin et al., 1964; Sarich and Wilson, 1966).

The immunological comparison of proteins is performed, in
outline, as follows.   A protein, say albumin, is purified from
an animal, say a chimpanzee.   The protein is injected onto a

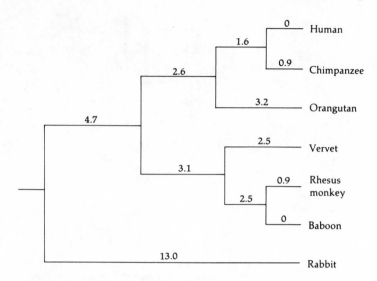

FIGURE 10.   *Phylogeny of various primates, based on the*
              sequence of 115 amino acids in carbonic
              anhydrase I.   The numbers on the branches are
              the estimated numbers of nucleotide substitutions
              that have occurred in evolution. (After Tashian
              et al., 1976).

rabbit or some other mammal, which develops an immunological
reaction and produces antibodies against the foreign protein
(antigen). The antibodies can be collected by bleeding the
rabbit; thereafter, they will react not only against the
specific antigen (chimpanzee albumin in the example), but
also against other related proteins (such as albumins from
other primates). The greater the similarity between the
protein used to immunize the rabbit and the protein tested,
the greater the extent of the immunological reaction. The
degrees of dissimilarity between the homologous proteins from
different species are expressed as "immunological distances."

     An efficient and sensitive immunological method that
requires only small amounts of purified protein is microcomplement fixation. "Complement" is a series of sequentially
acting chemical substances found in vertebrate serum. When
complement is added to antigens and antibodies under appropriate

experimental conditions, it becomes fixed within the three-dimensional lattice of the antigen-antibody complexes. The amount of antigen-antibody reaction is measured by determining the amount of complement fixed in the reaction. Suitably prepared ("sensitized") red blood cells are added, and any complement not fixed by the antibody-antigen complexes is available to lyse the cells. The amount of lysed cells is determined spectrophotometrically. The number of lysed red blood cells is proportional to the amount of free (unfixed) complement. The greater the amount of lysed cells the lesser the extent of the antigen-antibody reaction. Immunological distances measured by microcomplement fixation are approximately proportional to the number of amino acid differences of related proteins (Champion *et al.*, 1974).

Microcomplement fixation of various proteins has been used to determine immunological distances in a variety of organisms, including mammals, birds, and amphibians. Table 5 shows the immunological distances between man, apes, and Old World monkeys. Antibodies were prepared independently against albumin obtained from man (*Homo sapiens*), chimpanzee (*Pan troglodytes*) and gibbon (*Hylobates lar*). These were reacted against albumins obtained from seven species of apes and six species of Old World monkeys (*Cercopithecoidea: Macaca mulatta, Papio papio, Cercocebus galeritus, Cercopithecus aethiops, Colobus polykomos*, and *Presbytis entellus*). The albumins of the two species of chimpanzee appear identical. The tests with antiserum prepared against man show that albumins from the African apes (chimpanzee and gorilla) are more similar to human albumin than the albumins from the Asiatic apes (orangutan, siamang, and gibbon); albumins from the Old World monkeys are most different. The antiserum to chimpanzee yields similar results; chimpanzee albumin is more similar to the albumins

Table 5. Immunological distance between albumins of various
          Old World primates. (Calculated from data in
          Sarich and Wilson, 1967).

| | Antiserum to | | |
Species tested	Homo	Pan	Hylobates
Homo sapiens (human)	0	3.7	11.1
Pan troglodytes (chimpanzee)	5.7	0	14.6
Pan paniscus (pygmy chimpanzee)	5.7	0	14.6
Gorilla gorilla (gorilla)	3.7	6.8	11.7
Pongo pygmaeus (orangutan)	8.6	9.3	11.7
Symphalangus syndactylus			
(siamang)	11.4	9.7	2.9
Hylobates lar (gibbon)	10.7	9.7	0
Old World monkeys			
(average of six species)	38.6	34.6	36.0

of man and gorilla than to those of the Asian apes, while the
Old World monkeys are most different. The antiserum to gibbon's
albumin indicates that the albumins of gibbon and siamang are
very similar; it also indicates that orangutan's albumin is
not much more different from albumins of the other Asian apes
than the albumins of the African apes. A phylogenetic tree
based on the albumin distances is depicted in fig. 11.

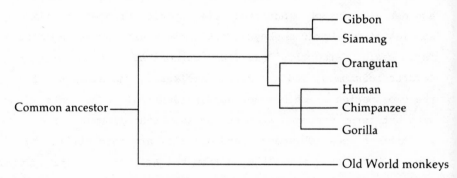

FIGURE 11.   Phylogeny of man, apes, and Old World monkeys,
             based on immunological differences between
             their albumin proteins. Man, the chimpanzee,
             and the gorilla appear more closely related to
             each other than either one of them is to the
             orangutan (After Sarich and Wilson, 1967.)

Lysozyme is an enzyme present in most animal species as well as in many plants and microorganisms. Immunological distances between lysozyme of man or baboon and a variety of organisms are shown in table 6; the primate lysozyme was obtained from milk. In comparison with human lysozyme, the lysozyme of chimpanzee appears to be identical (Immunological Distance, I.D. = 0), that of orangutan very similar (I.D. = 1), but that of gorilla quite different (I.D. = 32). These results are not consistent with those obtained with albumin, which indicate that gorilla is more similar to man than orangutan. Another anomaly in table 6 is that Old World monkeys appear more closely related to the New World monkeys than to man and the apes (see the last column in the table). Yet, there is ample evidence indicating that the phyletic line leading to the New World monkeys separated from the lineage leading to the Old World monkeys before the latter separated from the lineage leading to man and the apes. One more inconsistency in table 6 involves the immunological distance between man and baboon -- when anti-human serum is used this I.D. is 126, but the I.D. is 66 when anti-baboon serum is used.

The inconsistencies pointed out in the previous paragraph do not invalidate the use of microcomplement fixation or other immunological methods for evolutionary studies. Rather, the inference to be drawn is simply that phylogenies should be based on all available evidence, not just on one single trait. Similar inconsistencies occur in other kinds of studies. For example, if gross external morphology alone is considered, dolphins and seals might appear more closely related to some fishes than to terrestrial mammals. In order to establish phylogenetic relationships, data obtained from the immunological study of a given protein should be combined with immunological studies of other proteins, with other biochemical evidence,

Table 6.   Immunological distances between the lysozymes of man
           or baboon and those of various primates. (Data from
           Wilson and Prager, 1974.)

| Species tested | Anti-lysome to | |
	Homo	Papio
Homo sapiens *(man)*	0	66
Pan troglodytes *(chimpanzee)*	0	67
Pongo pygmaeus *(orangutan)*	1	76
Gorilla gorilla *(gorilla)*	32	38
Old World monkeys:		
Cercopithecus aethiops *(green monkey)*	93	6
Cercopithecus talapoin *(talapoin)*	114	3
Macaca mulatta *(rhesus)*	122	1
Macaca speciosa *(stump-tailed macaque)*	124	2
Macaca fascicularis *(crab-eating macaque)*	130	2
Macaca radiata *(bonnet macaque)*	131	2
Papio cynocephalus *(baboon)*	127	0
New World monkeys:		
Saimiri sciureus *(squirrel monkey)*	127	36
Saguinus oedipus *(tamarin)*	134	37
Callithrix jacchus *(marmoset)*	137	36

as well as with morphological, behavioral, and any other relevant
information.

ELECTROPHORETIC PHYLOGENIES

     Electrophoresis is a relatively inexpensive method used
to measure protein differences between organisms. With electrophoresis, the number of amino acid differences between two
species is not known, but only whether or not two proteins are
electrophoretically identical. The simplicity of the method
makes feasible the comparison of many proteins. The overall
results can be expressed as genetic distances, D (Ayala and
Kiger, 1980).

Electrophoresis is ineffective for comparing organisms that are evolutionarily very distant. These are likely to be electrophoretically different at all, or most, loci. Since the number of amino acid differences involved cannot be determined (but only whether or not the two proteins compared have identical migration), the method fails to discriminate the degree of differentiation between various species when these differ at all, or nearly all, loci. On the other hand, electrophoretic distances have the advantage of being based on many loci; therefore, unequal rates of evolution in different lineages with respect to one locus may be compensated by other loci. Electrophoresis is, in general, an appropriate method for measuring genetic change between closely related organisms, in which the amino acid sequence of a single protein may fail to show any differences or give misleading results because of the few numbers of substitutions involved.

The genetic distances between humans and apes based on the electrophoretic study of 23 proteins are shown in table 7 (Bruce and Ayala, 1979). The phylogeny reconstructed from the matrix of genetic distances is shown in fig. 12. The numbers given along the branches are expressed in units of genetic distance, D, and therefore estimate the average number of electrophoretically detectable allelic substitutions per locus that have occurred in each branch. The region where the great apes and humans branch from each other is shaded over, indicating that the differences observed in that region are much too small and not statistically significant. That is, these electrophoretic results show that the lineages of the great apes and humans branched from each other within a relatively short time span, but the results do not tell us in what sequence the lineages branched from one another.

Table 7. Genetic distance between pairs of nine taxa representing all six extant genera of hominoids. (From Bruce and Ayala, 1979).

	Homo	Pan t.	Pan p.	Gorilla	Pongo p.a.	Pongo p.p.	H. lar	H. concolor
Homo sapiens								
Pan troglodytes	.386							
Pan paniscus	.312	.103						
Gorilla gorilla	.373	.373	.385					
Pongo pygmaeus abelii	.347	.304	.238	.484				
Pongo pygmaeus pygmaeus	.350	.223	.115	.437	.130			
Hylobates lar	.716	.673	.716	.482	.598	.540		
Hylobates concolor	.847	.807	.847	.622	.592	.632	.130	
Symphalangus syndactylus	1.099	.793	1.099	.806	.756	.856	.337	.211

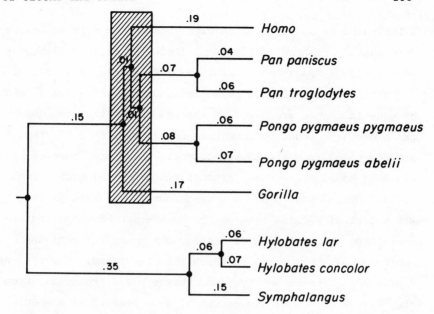

FIGURE 12. *Phylogeny of nine species and subspecies of hominoids based on data for 23 gene loci obtained by gel electrophoresis. The shading over the area where man and the great apes branch off from each other indicates that the difference in genetic distance between the four genera are too small to be reliable. (From Bruce and Ayala, 1979.)*

ORTHOLOGOUS AND PARALOGOUS GENES

Recent developments in molecular biology have made possible obtaining the nucleotide sequences of genes (DNA). Although the number of genes that have been sequenced is still relatively small, new sequences are being obtained at a fast rate. The nucleotide sequences of genes provide more evolutionary information than any of the techniques previously discussed. Indeed, it is in these sequences where the evolutionary and genetic information resides, whereas proteins are expressions of this information, and thus part of the conformation of organisms. The use of DNA sequences in evolutionary studies will be herein

illustrated by showing the results obtained by this technique
in the study of gene duplications (Leder *et al.*, 1980; Maniatis
et al., 1980; Proudfoot *et al.*, 1980; Efstratiadis *et al.*, 1981).

The reconstruction of phylogeny and the estimation of amount
of genetic change are based on the assumption that the genes, or
the proteins encoded by them, are *homologous*, that is, that they
are descended from a common ancestor. There are, however, two
kinds of homologous genes: orthologous and paralogous (Fitch,
1976). *Orthologous* genes are descendants of a gene present in
the ancestral species from which the species have evolved.
Therefore, the evolution of orthologous genes reflects the
evolution of the *species* in which they are found. The cytochromes
c of humans, rhesus monkeys, and horses are orthologous, because
they derive from a single ancestral gene present in a species
ancestral to the three organisms.

Paralogous genes are descendants of a *duplicated* ancestral
gene. Paralogous genes, therefore, evolve within the same
species (as well as in different species). The genes coding
for the α, β, γ, and δ hemoglobin chains in humans are paralogous
The evolution of paralogous genes reflects differences that have
accumulated since the genes duplicated. Homologies between
paralogous genes serve to establish *gene* phylogenies, *i.e.*, the
evolutionary history of duplicated genes within a given lineage.

The amino acid sequences of the α, β, γ, and δ hemoglobin
chains and of myoglobin, a closely related protein, are known
in humans as well as in other organisms. These sequences have
made possible to reconstruct the evolutionary history of the
duplications that gave rise to the corresponding genes (fig. 13).
But the direct study of the nucleotide sequences of these genes
has shown that the situation is more complex, and also more
interesting, than appears from the protein sequences.

Until very recently, it was assumed that the triplets

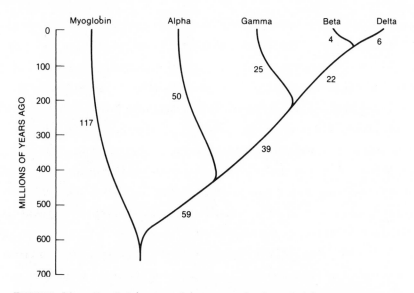

FIGURE 13.   *Evolutionary history of the globin genes based
on the amino acid sequences of the encoded
proteins. Oxygen storage and transport is
mediated by myoglobin in muscle and by hemoglobin in blood. The branching points indicate
where the ancestral genes were duplicated
giving rise to a new gene line. The minimum
number of nucleotide replacements required to
account for the amino acid differences between the
proteins are indicated along the branches. The
first gene duplication occurred about 600 million
years ago; one gene coding for myoglobin and
the other being the ancestor of the various
hemoglobin genes. Around 400 million years ago,
the hemoglobin gene became duplicated into one
leading to the modern alpha gene and another
which would duplicate again around 200 million
years ago into the gamma and the beta genes.
The beta gene duplicated again some 40 million
years ago in the ancestral lineage of the
higher primates, giving rise to one new gene
coding for the delta hemoglobin chain.*

coding for the amino acids of a given polypeptide were all

contiguous. DNA sequencing has shown that, in eukaryotes, the

coding sequences of most genes are interrupted, broken into two or more segments separated by nucleotide sequences that do not code for amino acids (fig. 14). The segments coding for amino acids are called *exons*, the intervening sequences are called *introns* (Chambon, 1981). The number of introns varies from gene to gene. A gene coding for a collagen has 52 introns; a gene coding vitellogenin has 33 introns; the genes coding for the hemoglobin polypeptides consist each of two introns, which separate the three exons that encóde the amino acids. Some genes, such as those coding for histone proteins and for interferon, have no introns, but they seem to be the exception rather than the rule.

Several steps are required for the formation of the mature messenger RNA that eventually becomes associated with ribosomes and serves as a template for translation (fig. 15). First, a primary transcript is formed: a messenger RNA that transcribes the complete DNA sequence which makes up the gene, the exons and the intervening introns. (This primary transcript contains the flanking sequences that exist at both ends of the gene and

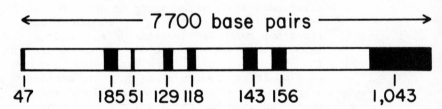

FIGURE 14.   Schematic representation of the gene for pro-
              tein ovoalbumin showing the split organization
              characteristic of most eukaryotic genes. This
              gene consists of eight exons (black) and seven
              introns (white). The exons contain the
              nucleotide sequences coding for the amino acids
              in the protein. The number of nucleotide
              base pairs in each exon are indicated (After
              Chambon, 1981).

that will remain an integral part of the mature messenger RNA,
although they are not translated into amino acids). Then, the
introns are excised in several splicing steps performed by
special enzymes (endonucleases). These steps are accomplished
in the cell nucleus. The final product is the mature messenger
RNA, which migrates from the nucleus to the cytoplasm, where
it becomes associated with ribosomes; translation ensues
(Chambon, 1981).

It has now become known that the number of hemoglobin
genes is greater than previously thought. Table 8 gives a
summary of the various kinds of hemoglobins known in humans and

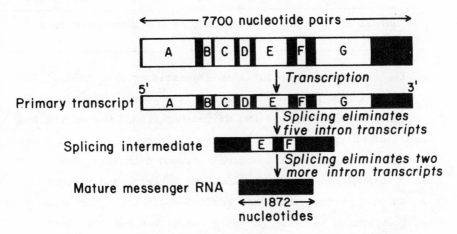

FIGURE 15.   *Schematic representation of the production of
the mature messenger RNA from the ovoalbumin
gene. First the entire gene is transcribed
into a precursor RNA. (The transcript is
"capped" at the 5' end and a "tail" of adenine
nucleotides is added to the 3' end; these are
not illustrated.) Then the transcripts of the
introns are excised and the adjacent transcripts are ligated. This splicing occurs in
a number of steps; only one of the inter-
mediates, with two remaining introns, is
shown. These steps occur in the nucleus; the
mature messenger is then transferred to the
cytoplasm where translation takes place.
(After Chambon, 1981).*

Table 8.   The hemoglobins present at different stages of human
           life, their tetramer structure, and the genes
           encoding the corresponding polypeptides.

Hemoglobin		Tetramer Structure	Genes
Adult*	A	$\alpha_2 \beta_2$	$\alpha 1$ and $\alpha 2$, β
	A_2	$\alpha_2 \delta_2$	$\alpha 1$ and $\alpha 2$, δ
Fetus	F	$\alpha_2 \gamma_2$	$\alpha 1$ and $\alpha 2$, $^G\gamma$ and $^A\gamma$
Embryo (to 8 weeks):	Gower 1	$\alpha_2 \varepsilon_2$	$\alpha 1$ and $\alpha 2$, ε
Transition from embryo to fetus:	Gower 2	$\alpha_2 \varepsilon_2$	$\alpha 1$ and $\alpha 2$, ε
	Portland	$\zeta_2 \gamma_2$	$\zeta 1$ and $\zeta 2$, $^G\gamma$ and $^A\gamma$

*
In normal adults, about 98 percent of the hemoglobin is A,
whereas the remaining 2 percent is A_2.

of the genes that code for them (Maniatis et al., 1980).   The
functional hemoglobins are tetramers, consisting of two polypeptides of one kind and two of another kind.   One of the two
kinds of polypeptide is ε in embryonic hemoglobin, γ in fetal
hemoglobin, β in adult hemoglobin A_1 and δ in adult hemoglobin
A_2.   (Hemoglobin A makes up about 98 percent, and hemoglobin
A_2 about 2 percent, of human adult hemoglobin).   The other kind
of polypeptide is ζ in embryonic hemoglobin and α in fetal or
adult hemoglobin.   The genes coding for one kind of polypeptide
(ε, γ, β, and δ) are located on chromosome 11; the genes coding
for the second kind of polypeptide are located on chromosome 16
(fig. 16).

But there are additional complexities.   Two γ genes exist,
known as $^G\gamma$ and $^A\gamma$, and there also exist two ζ genes ($\zeta 1$ and
$\zeta 2$) as well as two α genes ($\alpha 1$ and $\alpha 2$).   In addition, there are
two β pseudogenes ($\psi\beta 1$ and $\psi\beta 2$) and one α pseudogene ($\psi\alpha 1$).
These pseudogenes are not translated and probably not transcribed
either.   The pseudogenes are very similar in nucleotide sequence

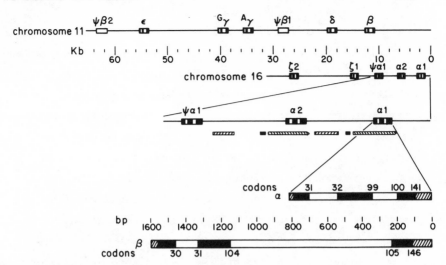

FIGURE 16. *Chromosomal arrangement of the globin genes in humans. Top: the β-like genes are on chromosome 11; the α-like genes on chromosome 16. Middle: the α1 and α2 genes are identical in nucleotide sequence; in addition, the nucleotide sequence between ψα1 and α2 is largely similar to the sequence between α2 and α1. The corresponding regions of homology are indicated by identical shading. Bottom: The α and β genes, as well as all the other globin genes, have the same organization: three exons (black) separated by two introns (white). The length of the introns is variable, but their position is the same -- they are located between functionally identical codons (e.g., codon 30 of β corresponds to codon 31 of α, because the β gene has one fewer codon than the α gene in the early part of the sequence; both codons code for arginine); the conservation of the number and position of the introns throughout the hundreds of thousands of years of evolution of the globin genes indicates that the introns play a functional role. The scale on top is in kilobases (Kb) or thousands of base pairs; the scale on bottom is in number of base pairs (bp).*

to the corresponding functional genes, but they include
terminating codons and other mutations that make it impossible
for them to yield functional hemoglobins. It is tempting to
speculate that pseudogenes may represent intermediate, nonfunctional stages of genes that are allowed freely to mutate
and that may eventually evolve into genes with functions
somewhat different from those of the ancestral genes from
which they arose by duplication. That is, it may be that
some of the presently functional genes coding for various
hemoglobin chains may have evolved from ancestral sequences
that some time in the past were nonfunctional pseudogenes.

The similarity in the nucleotide sequence of the globin
genes, and pseudogenes, of both the α and β gene families
indicates that they are all homologous, *i.e.*, that they have
arisen through various duplications and subsequent evolution
from a gene ancestral to all. Moreover, homology exists also
between the nucleotide sequences that separate one gene from
another. For example, the nucleotide sequence that separates
$\psi\alpha1$ and $\alpha2$ is largely similar with the sequence that separates
$\alpha2$ from $\alpha1$. The segments that are clearly homologous are
indicated in the middle of fig. 16 by the bars with similar
shading underlying these chromosomal regions. The different
size of the gaps that separate the bars bring up an additional
point; namely, that additions and deletions of nucleotides
occur that change the size of a given DNA segment. This was
known for the coding parts of genes; for example, the α
hemoglobin chain consists of 141 amino acids whereas the β
hemoglobin chain consists of 146 amino acids. Not surprisingly,
additions and deletions of single or few nucleotides also occur
in the noncoding parts of the genome -- in the sequences that
separate genes, as already pointed out, and in the introns
as well, as it is apparent by comparing their different lengths
in the α and β genes (see bottom of fig. 16). At least some

of these additions arise by duplication of very short nucleotide sequences, something manifested by the identification
within a given intron (as well as in other parts of the genome)
of short sequences present several times either in tandem or
interspersed with other sequences.

It thus appears that DNA duplications may be of variable
magnitude. Some duplications involve only a few nucleotides,
as those just pointed out. Others involve a complete gene and
these may also include the adjacent noncoding sequences. Still
other duplications involve large chromosomal regions, including
more than one gene -- the similarity of arrangement between the
α gene family in chromosome 16 and the β gene family in
chromosome 11 supports this last inference.

It was pointed out above that different proteins (and
the genes coding for them) evolve at different rates: cytochrome *c* evolves slowly, the fibrinopeptides evolve rapidly
(fig. 9). We now know that, moreover, exons evolve at
different rates than introns, that some parts of introns
evolve faster than others, and that introns and exons evolve
at rates different from those of the sequences that disjoin
one gene from another (Leder *et al.*, 1980; Zimmer *et al.*,
1980). The DNA of an organism is like a large set of clocks,
each ticking at a different rate, but all timing the same
evolutionary events (Fitch, 1976).

THE NEUTRALITY THEORY OF MOLECULAR EVOLUTION

The reconstruction of phylogeny from genetic similarities
depends on the assumption that degrees of similarity reflect
degrees of phylogenetic propinquity. On the whole, this is
a reasonable assumption because evolution is a process of
gradual change. However, differences in rates of genetic

change among lineages may be a source of error. Assume that
a species, *A*, diverged from the common lineage of two other
species, *B* and *C*, before the latter diverged from each other.
Assume that a certain protein has evolved at a much faster rate
in the lineage leading to *C* than in the other two lineages. It
might be the case that the amino acid sequence of the protein
would be more similar between *A* and *B* than between *B* and *C*.
The phylogeny inferred from the sequence of the protein might
be erroneous.

The hypothesis has recently been advanced by Motoo Kimura
and others (Kimura, 1968; Kimura and Ohta, 1971; King and Jukes,
1969) that rates of amino acid replacements in proteins and of
nucleotide substitutions in DNA may be approximately constant
because the vast majority of such changes are selectively
neutral. New alleles appear in a population by mutation. If
alternative alleles have identical fitness, changes in allelic
frequencies from generation to generation would occur only by
accidental sampling errors from generation to generation -- by
genetic drift. Rates of allelic substitution would, then, be
"stochastically" constant, *i.e.*, would occur with a constant
probability for a given protein. That probability can be shown
to be simply the mutation rate for neutral alleles.

The neutrality theory of molecular evolution admits that,
for any gene, a large proportion of all possible mutants are
harmful to their carriers; these mutants are eliminated or kept
at very low frequency by natural selection. The evolution of
morphological, behavioral, and ecological traits is largely
governed by natural selection, because it is determined by the
selection of favorable mutants against deleterious ones. It is
assumed, however, that a number of favorable mutants, adaptively
equivalent to each other, can occur at each locus. These
mutants are not subject to selection relative to one another

because they do not affect the fitness of their carriers (nor modify their morphological, physiological, or behavioral properties). According to the neutrality theory, evolution at the molecular level consists for the most part of the gradual replacement of one neutral allele by another one, functionally equivalent to the first. The theory assumes that, although favorable mutations occur, they are so rare that they have little effect on the overall evolutionary rate of nucleotide and amino acid substitutions (Ayala, 1977).

Neutral alleles are not defined as having fitnesses identical in the mathematical sense. Operationally, neutral alleles are those whose differential contributions to fitness are so small that their frequencies change more owing to drift than to natural selection. Assume that two alleles, A_1 and A_2, have fitnesses 1 and $1 - s$ (where s is a positive number smaller than 1). The two alleles are effectively neutral if, and only if

$$4N_e s < < 1$$

where N_e is the effective size of the population.

We now want to find the rate of substitution of neutral alleles, k, per unit time in the course of evolution. Time units can be years or generations. In a random mating with N diploid individuals

$$k = 2Nux$$

where u is the neutral mutation rate per gamete per unit time (time measured in the same units as for k), and x is the probability of ultimate fixation of a neutral mutant. The derivation of the equation above is straight forward -- there are $2Nu$ mutants per unit time, each with a probability x of becoming fixed (Kimura, 1968).

A population of N diploid individuals has $2N$ genes at each locus. If the alleles are neutral, all genes have

identical probability of becoming fixed; this probability is
simply

$$x = \frac{1}{2N}$$

Replacing the value of x in the previous equation, we obtain

$$k = 2Nu\left(\frac{1}{2N}\right) = u$$

That is, the rate of substitution of neutral alleles is
precisely the rate at which the neutral alleles arise by
mutation, independently of the size of the population and
any other parameters. This is not only a remarkably simple
result, but also one with momentous implications if it indeed
applies to molecular evolution.

THE MOLECULAR CLOCK OF EVOLUTION

   If the neutrality theory of molecular evolution were
correct for a large number of gene loci, protein and DNA
evolution would serve as evolutionary clocks. The degree of
genetic differentiation between species would be a measure of
their phylogenetic relatedness; it would be thus justified to
reconstruct phylogenies on the basis of genetic differences.
Moreover, the actual "chronological" time of the various
phylogenetic events could be approximately estimated. Assume
that we have a phylogeny such as the one shown in fig. 8.
If the rate of evolution of cytochrome c were constant through
time, the number of nucleotide substitutions that have occurred
in each branch of the phylogeny would be directly proportional
to the time elapsed. If we know from an outside source (such
as the paleontological record) the actual geological time of
any one event in the phylogeny, it becomes possible to determine
the times of all other events by a simple proportion. That is,
once it is "calibrated" by reference to one single event, the

molecular clock can be used to measure the time of occurrence
of all other events in a phylogeny.

The molecular clock postulated by the neutrality theory
is, of course, not a "metronomic clock," like timepieces in
ordinary life that measure time exactly. The neutrality
theory predicts, instead, that molecular evolution is a
"stochastic clock," like radioactive decay (Fitch, 1976). The
probability of change is constant, although some variation
occurs. Over fairly long periods of time a stochastic clock
is, nevertheless, quite accurate. Moreover, each gene or
protein would be a separate clock, providing an independent
estimate of phylogenetic events and their time of occurrence.
Each gene or protein would "tick" at a different rate (the
mutation rate to neutral alleles, u, of the gene; see fig. 9)
but all genes would be timing the same evolutionary events.
The joint results of several genes or proteins would provide
a fairly precise evolutionary clock.

Is there a molecular clock of evolution? This question
can be investigated by examining whether or not the variation
in the number of molecular changes occurred during equal
evolutionary periods is greater than expected by chance. This
would also be a test of the neutrality theory of molecular
evolution. Two kinds of test are possible. One kind of test
consists of examining the number of molecular changes between
phylogenetic events whose timing is known from the paleontological record and other sources. The other kind of test
does not use actual times, but rather looks at parallel lineages
derived from a common ancestor and tests whether the variation
in the number of molecular changes along the branches is greater
than expected by chance.

Whether or not the neutrality theory is correct, and how
accurate the molecular clock is, are at present controversial

matters. The existing evidence suggests that the variation
in the rate of molecular evolution is greater than predicted
by the neutrality theory. Nevertheless, molecular evolution
appears to occur with sufficient regularity to serve as an
evolutionary clock, although not as accurate as if the rate
of evolution were stochastically constant with the variation
expected from a Poisson distribution. The results of one test
devised by Charles H. Langley and Walter M. Fitch (1974) are
given in table 9.

The test uses seven proteins sequenced in 17 mammals and
starts by adding up the proteins one after another and treating
them as if they were one single sequence. The minimum number of
nucleotide substitutions is found that accounts for the descent
of the amino acid sequences from a common ancestor; the numbers
of substitutions are then assigned to the various branches in
the phylogeny. Two independent tests are made. First, the
total number of substitutions per unit time is examined for
different times; the hypothesis tested is whether the *overall*
rate of change is uniform over time. The probability that

Table 9. *Statistical tests of the constancy of evolutionary*
          rates of seven proteins in 17 species of mammals.
          (Data from Langley and Fitch, 1974)

	Chi-square	Degrees of freedom	Prob-ability
Overall rates (comparisons among branches over all seven proteins)	82.4	31	4×10^{-6}
Relative rates (comparisons among proteins within branches)	166.3	123	6×10^{-2}
Total	248.7	154	6×10^{-6}

the variation observed is due to chance, is 4×10^{-6}, statistically highly significant. The conclusion follows that the proteins have not evolved at a constant rate with a Poisson variance. It is possible, however, that the proteins have all changed their rates *proportionately,* for example because the rate of molecular evolution is constant per generation rather than per year; variations in generation length might have occurred through time. This possibility is tested by examining whether the rates of evolution of one protein *relative* to another are uniform through time. There is a marginally significant deviation from expectation (probability $\simeq 0.06$). The probability that all the variation observed (*total*) is due to chance is extremely small, 6×10^{-6}.

The test is particularly valid because it makes no use of paleontological dates. The phylogeny is constructed using the protein data alone and, thus, maximizes the probability of agreement between the data and the hypothesis of stochastically constant rates of molecular evolution. Even so, the data do not fit the hypothesis of constant probability of change with a Poisson variance. John H. Gillespie and C.H. Langley (1979) have recently shown that the data used for table 9 are consistent with the hypothesis that molecular evolution occurs with a constant probability, if it is assumed that the variance of this probability is greater than expected from a Poisson distribution. Nevertheless other statistical analyses of protein sequences have confirmed the conclusion of Langley and Fitch (1974) that proteins do not evolve according to the prediction of the neutrality theory (Corruccini *et al.,* 1980). This conclusion has been further confirmed by the study of DNA sequences: not even synonymous codons (coding for identical amino acids) are evolving at random (Modiano *et al., 1981),* but rather there are functional constraints against synonymous codon changes in functional genes (Miyata and Yasunaga, 1981).

It thus appears that molecular evolution is not stochastically constant, but it is also the case that the heterogeneity of evolutionary rates is not excessively large. It is
therefore, possible to use genetic data as an approximate
evolutionary clock, although in order to avoid large errors it
is necessary to use *average* rates obtained for many proteins
and for long periods of time. Fig. 17 plots the cumulative
number of nucleotide substitutions required in seven proteins
against the paleontological dates of divergence in the evolution
of 16 mammal species. The overall correlation is fairly good
for all phylogenetic events except those involving some primates
which appear to have evolved at a substantially lower rate than
the average. This deviation from the average, observed at the
lower left of the figure, illustrates an important point --
the more recent divergence of any two species the more likely
it is that the genetic changes observed will depart from the
average evolutionary rate. This is simply because, as time
increases, periods of rapid evolution and periods of slow
evolution in any one lineage will tend to cancel out each
other.

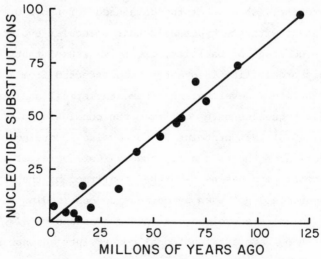

REFERENCES CITED

AYALA, F.J. 1977. Protein evolution in different species:
Is it a random process? *In:* Molecular Evolution and
Polymorphisms. M. Kimura, ed. Japan, Mishima: pp. 93-102.
AYALA, F.J. and J.A. KIGER. 1980. Modern Genetics. Menlo
Park, California: Benjamin/Cummings.
BRUCE, E.J. and F.J. AYALA. 1979. Phylogenetic relationships
between Man and the Apes: Electrophoretic evidence.
Evolution 33:1040-1056.
CHAMBON, P. 1981. Split genes. Scientific American 244:60-71
(May).
CHAMPION, A.B., E.M. PRAGER, D. WACHTER, and A.C. WILSON. 1974.
Microcomplement fixation. *In:* Biochemical and Immunological
Taxonomy of Animals. C.A. Wright, ed., pp. 397-416. London:
Academic Press.
CORRUCCINI, R.S., M. BABA, M. GOODMAN, R.L. CIOCHON, and
J.E. CRONIN. 1980. Non-linear macromolecular evolution
and the molecular clock. Evolution 34:1216-1219.
DICKERSON, R.E. 1971. The structure of cytochrome *c* and the
rates of molecular evolution. J. Molec. Evol. 1:26-45.
DOBZHANSKY, TH., F.J. AYALA, G.L. STEBBINS, and J.W. VALENTINE.
1977. Evolution. San Francisco: Freeman.

---

*FIGURE 17. Nucleotide substitutions versus paleon-
tological time. The total nucleotide substitutions for seven proteins (cytochrome* c,
*fibrinopeptides A and B, hemoglobins alpha
and beta, myoglobin, and insulin c-peptide)
have been calculated for comparisons between
pairs of species whose ancestors diverged at
the time indicated in the abscissa. The
solid line has been drawn from the origin to
the outermost point, and corresponds to a
total rate of 0.41 nucleotide substitutions
per million years (or 98.2 nucleotide substitutions per 2 x 120 million years of
evolution) for the genes coding for all seven
proteins. The fit between the observed number
of nucleotide substitutions and the expected
number (as determined by the solid line) is
fairly good in general. However, in the
primates (points below the diagonal at lower
left) protein evolution seems to have occurred
at a slower rate than in most other organisms.
(After Fitch, 1976).*

EFSTRATIADIS, A., J.W. POSAKONY, T. MANIATIS, R.M. LAWN, C.
O'CONNELL, R.A. SPRITZ, J.K. DeRIEL, B. FORGET, S.M.
WEISSMAN, J.L. SLIGHTOM, A.E. BLECHL, O. SMITHIES, F.E.
BARALLE, C.C. SHOULDERS, and N.J. PROUDFOOT. 1981. The
structure and evolution of the human β-globin gene family.
Cell 21:653-668.

FITCH, W.M. 1976. Molecular evolutionary clocks. *In:*
Molecular Evolution. F.J. Ayala, ed. Sinauer Assoc.,
pp. 160-178.

FITCH, W.M., and E. MARGOLIASH. 1967. Construction of phylogenetic trees. Science 155:279-284.

FITCH, W.M., and E. MARGOLIASH. 1970. The usefulness of
amino acid and nucleotide sequences in evolutionary
studies. Evol. Biol. 4:67-109.

GILLESPIE, J.H., and C.H. LANGLEY. 1979. Are evolutionary
rates really variable? J. Mol. Evol. 13:27-34.

HOYER, B.H., B.J. McCARTHY, and E.T. BOLTON. 1964. A molecular
approach in the systematics of higher organisms. Science
144:959-967.

KENDREW, J.C. 1968. Information and conformation in biology.
In: Structural Chemistry and Molecular Biology. J.C.
Kendrew, A. Rich, and N. Davidson, eds. San Francisco:
Freeman, Pp. 187-197.

KIMURA, M. 1968. Evolutionary rate at the molecular level.
Nature. 217:624-626.

KIMURA, M., and T. OHTA. 1971. Protein polymorphism as a
phase of molecular evolution. Nature 229:467-469.

KING, J.L., and T.H. JUKES. 1969. Non-Darwinian evolution.
Science 164:788-798.

KOHNE, D.E., J.A. CHISCON, and B.H. HOYER. 1972. Evolution
of primate DNA sequences. J. Human Evol. 1:627-644.

LAIRD, C.D., and B.J. McCARTHY. 1968. Magnitude of interspecific nucleotide sequence variability in *Drosophila*.
Genetics 60:303-322.

LAIRD, C.D., B.L. McCONAUGHY, and B.J. McCARTHY. 1969. Rate
of fixation of nucleotide substitutions in evolution.
Nature 224:149-154.

LANGLEY, C.H., and W.M. FITCH. 1974. An examination of the
constancy of the rate of molecular evolution. J. Mol.
Evol. 3:161-177.

LEDER, P., J.N. HANSEN, D. KONKEL, A. LEDER, Y. NISHIOKA, and
C. TALKINGTON. 1980. Mouse globin system: A functional
and evolutionary analysis. Science 209:1336-1342.

MANIATIS, T., R.C. HARDISON, E. LACY, J. LAUER, C. O'CONNELL,
D. QUON, G.K. SIM, and A. EFSTRATIADIS. 1978. The
isolation of structural genes from libraries of eucaryotic
DNA. Cell 15:687-701.

MANIATIS, T., E.F. FRITSCH, J. LAUER, and R.M. LAWN. 1980.
The molecular genetics of human hemoglobins. Ann. Rev.
Genet. 14:145-178.

McCARTHY, B.J., and M.N. FARQUHAR. 1974. The rate of change
of DNA in evolution. *In:* Evolution of Genetics Systems.
Brookhaven Symp. Biol. No. 23, pp. 1-41.

MIYATA, T., and T. YASUNAGA. 1981. Rapidly evolving mouse
α-globin-related pseudo gene and its evolutionary history.
PNAS 78:450-453.

MODIANO, G., G. BATTISTUZZI, and A.G. MOTULSKY. 1981. Non-
random patterns of codon usage and of nucleotide substitutions in human α- and β-globin genes: An evolutionary
strategy reducing the rate of mutations with drastic
effects? PNAS 78:1110-1114.

PROUDFOOT, N.J., M.H.M. SHANDER, J.L. MANLEY, M.L. GEFTER, and
T. MANIATIS. 1980. Structure and in vitro transcription
of human globin genes. Science 209:1329-1336.

REICHLIN, M., M. HAY, and L. LEVINE. 1964. Antibodies to
human A_1 hemoglobin and their reaction with A_2, S, C, and
H hemoglobins. Immunochemistry 1:21-30.

SANGER, F., A.R. COULSON, T. FRIEDMANN, G.M. AIR, B.G. BARRELL,
N.L. BROWN, J.C. FIDDES, C.A. HUTCHISON III, P.M. SLOCOMBE,
and M. SMITH. 1978. The nucleotide sequence of bacterio-
phage φX 174. J. Molec. Biol. 125:225-246.

SARICH, V.M., and A.C. WILSON. 1966. Quantitative immunochemistry and the evolution of primate albumins: Microcomplement fixation. Science 154:1563-1566.

SARICH, V.M., and A.C. WILSON. 1967. Immunological time scale
for hominid evolution. Science 158:1200-1203.

SIMPSON, G.G. 1953. The Major Features of Evolution. New
York: Columbia University Press.

TASHIAN, R.E., M. GOODMAN, R.E. FERRELL, AND R.J. TANIS. 1976.
Evolution of carbonic anhydrase in primates and other
mammals. *In:* Molecular Anthropology. M. Goodman and
R.E. Tashian, eds. New York: Plenum.

WILSON, A.C., and E.M. PRAGER. 1974. Antigenic comparison
of animal lysozymes. *In:* Lysozyme. E.F. Ossermen,
R.E. Canfield, and S. Beychock, eds., pp. 127-141. New
York: Academic Press.

ZIMMER, E.A., S.L. MARTIN, S.M. BEVERLY, Y.W. KAN, and A.C.
WILSON. 1980. Rapid duplication and loss of genes
coding for the α chains of hemoglobin. PNAS 77:2158-2162.

AUTHOR INDEX

A

Adelburg, E.A.
  see Stanier et al
  1976, 16
Ainsworth, G.C. 16
Air, G.M.
  see Sanger et al
  1978, 266
Alexander, R.D. 177,178
Allaway, W.G. 125
Alston, R.E. 56
Altman, P.L. 16
Anderson, D.J.
  see Osmond et al
  1980, 100,103
Andrews, H.N.
  see Schopf et al
  1966, 33
Arends, T.
  see Weitkamp et al
  1972, 229
Arms, K. 154
Arnold, S.J. 6,176,191,
  193,194,195,197,200,
  205
Axelrod, D.I. 56
Ayala, F.J. 5,214,257,
  280,281,282,283,293
  see Dobzhansky et al
  1977, 260,266
Ayres, M.
  see Gershowitz et al
  1972, 229

B

Baba, M.
  see Corruccini et al

1980, 297
Baker, H.G. 7,133,134,135,
  136,138,139,140,141,142,
  143,146,147,149,150,151,
  153,154,155,156,157,160,
  161,162,163,165,166
Baker, I. 7,133,134,135,
  136,138,139,140,141,142,
  143,146,147,149,150,151,
  153,154,155,156,157,160,
  161,162,166
Balows, A.
  see Starr et al
  1981, 16
Banks, H.P. 32,42
Baralle, F.E.
  see Efstratiadis et al
  1981, 284
Barbel, E.R.
  see Mansell et al
  1976, 72
Barrell, B.G.
  see Sanger et al
  1978, 266
Barth, R.H. 177
Bassett, M.G.
  see Edwards et al
  1979, 32
Batanouny, K.H.
  see Ziegler 1981, 103
Bate-Smith, E.C. 57,60
Battersby, A.R. 66
Battistuzzi, G.
  see Modiano et al
  1981, 297
Bell, D.J. 187,204
Bell, E.A. 56

303

SUBJECT INDEX

A

Absorbing agents, ultra-
violet 70
Acanthaceae 104
Acer 58
Acid metabolism of
Crassulaceans 93ff
Adaptation, equilibrium 2
Agavaceae 106
Aizoaceae 104,106
Åland island populations
228,230
Alga(e) 70
green 51,75
marine 75
Algal plexus, ancestral
green 71
Alkaline hydrolysis 55
Alkaloids 67,68
benzylisoquinoline 66
nectar 163
Allelochemic(s) 68,78
activity 73
Allelochemicals 69,81
Amaranthaceae 104
American Indians
heterozygosity in 226ff
Amino acids 7,62,290
complements 155ff
concentrations 150ff
differences 264
nectar 150
occurrence 157
sequences 285
Ammonia-lyase(s) 64
phenylalanine 71
Anagenesis 261
Anagenetic change 263,265
evolution 262

B

Angiosperm(s) 32,56ff,80,82ff
compression fossils 58
Devonian vascular 79
fossils 32,56
herbs 81
Miocene 83
shrubs 81
Animalia 15,19,21,23,24
Ant repellents and nectar
165
Apiaceae 81
Apocynaceae 106
Apoplastic transport 75
Araucaria 81
Araucarioxylon 36
Arborescence 79
Archaebacteria 15
Archaeopteris 52
Arthropods, terrestrial 79
Asclepiadaceae 106
Ascomycotes 21
Associations, hereditary 11
level of 12
Asteraceae 104,106
Auxin 71,72,73

B

Bacteria 14,15,17
Banded tubular cells 32
Bantu peoples 226,229,232
Basidiomycotes 21
Bat-pollinated flowers 143
Bataceae 106
Benzylisoquinoline alkaloids
66
Biflavonyls 81
Biochemical diversification
80

315